T0345288

Intelligent Fatigue Statistics

This book introduces the principles and theories of fatigue statistics and presents the Zhentong Gao method along with its practical applications.

Fatigue in metal components is a complex phenomenon and an important issue because the fatigue life of key components often determines the overall lifespan of the aircraft. This book provides an overview and analysis of existing fatigue life calculation and estimation methods, including probability statistical methods such as using Gaussian distribution, logarithmic Gaussian distribution, and Weibull distribution to fit actual data. However, the authors argue that these methods have limitations in terms of reliability and applicability. They point out that the location parameter of the Weibull distribution represents a safe life with 100% reliability. Based on these methods, this book proposes the intelligent Zhentong Gao method, which differs from traditional approaches and effectively addresses and solves the problem of estimating the three parameters of Weibull distribution.

This book will serve as a valuable reference for researchers, senior undergraduate and graduate students, and engineers in the fields of aerospace, automotive, mechanical, civil engineering, etc.

Zhentong Gao, an academician of the Chinese Academy of Sciences, is an expert in metal structural fatigue. Before retiring, he worked as a professor at Beihang University, China, where he focused on the research of material fatigue performance and metal structural fatigue.

Jiajin Xu is a researcher specializing in the fields of semiconductors, laser physics, nonlinear optics, and computational physics. He worked as a visiting scholar at the Optical Research Center, University of Arizona, United States, between 1988 and 1991.

Intelligent Fatigue Statistics

Zhentong Gao and Jiajin Xu

CRC Press
Taylor & Francis Group
Boca Raton London New York

CRC Press is an imprint of the
Taylor & Francis Group, an **informa** business

Designed cover image: by the authors

First published in English 2025
by CRC Press
2385 NW Executive Center Drive, Suite 320, Boca Raton FL 33431

and by CRC Press
4 Park Square, Milton Park, Abingdon, Oxon, OX14 4RN

CRC Press is an imprint of Taylor & Francis Group, LLC

© 2025 Zhentong Gao and Jiajin Xu

Translated by Jiajin Xu

ISBN: 978-1-032-77309-4 (hbk)
ISBN: 978-1-032-78559-2 (pbk)
ISBN: 978-1-003-48847-7 (ebk)

DOI: 10.1201/9781003488477

Typeset in Minion
by codeMantra

Contents

Abbreviations

CDF	Cumulative distribution function
e.g.	exempli gratia (Latin language), for example
etc.	et cetera (Latin language), and so forth
GD	Gaussian distribution
G_ZT Gao method	Generalized Zhentong Gao method
IDE	Integrated development environment
i.e.	id est (Latin language), that is
IFS	Intelligent fatigue statistics
MGF	Moment generating function
MLE	Maximum likelihood estimation
LSE	Least square estimate
PDF	Probability density function
q.e.d.	quod erat demonstrandum (Latin language), proven
SD	Standard deviation
SND	Standard normal distribution
SSE	Sum of square for error
SSR	Sum of square for regression
SST	Sum of square for total
WD	Weibull distribution
ZT Gao method	Zhentong Gao method

Symbols

b	Shape parameter
N	Fatigue life
$\mathbf{N_0}$	Save life
n	Size of sample
P	Reliability (level)
$\mathbf{R^2}$	Determination coefficient
r	Correlation coefficient
$\mathbf{S^2}$	Deviation of sample
s	Standard deviation of sample
X̃	Average of sample
$\mathbf{x_0}$	Location parameter
α	Significance (level)
γ	Confidence (level)
λ	Scale parameter
μ	Mean of population
σ	Standard deviation of population

Foreword to the Chinese Edition

FATIGUE DAMAGE IS ONE of the main forms of structural failure, and the assessment of structural fatigue life has always been the focus of attention and technical difficulty in the field of aviation.

In 1986, Academician Gao Zhentong, the first author of this book, made a series of major original innovations, such as the two-dimensional dynamic stress-strength interference model based on summarizing the research results of fatigue statistics at home and abroad, and published "Fatigue Applied Statistics". This book created a new branch of the discipline and laid the theoretical foundation to determine the reliability and life of the fleet through the full-scale test data of an aircraft in China. After Mr. Gao's structural fatigue reliability life extension theoretical system was proposed, it not only doubled the life of fighter aircraft in service at that time but also continued to guide the life extension work of more than 20 aircraft models in my country, creating tens of billions of economic benefits, as well as national defense and social value that cannot be measured in money.

With the progress of calculation, analysis, and testing technologies, the assessment of structural fatigue life is evolving toward greater refinement and intelligence. Despite being 94 years old, Mr. Gao maintains a spirit of continuous innovation and self-breakthrough. He constantly concerns himself with the development of aircraft structural fatigue extension and delay technology in China and continuously guides students in their exploration. Mr. Xu Jiajin, a student of Mr. Gao more than 50 years ago, has benefited from Mr. Gao's tireless teachings even after more than 50 years since graduation. Finally, today, we present *Intelligent Fatigue Statistics* with the guidance of Mr. Gao.

This book is a significant development of "Fatigue Applied Statistics". It not only introduces modern intelligent analytical techniques into the traditional statistical analysis methods of fatigue problems but also, more importantly, proposes the "Zhentong Gao method" to estimate the three parameters of the Weibull distribution. This is a crucial development in traditional theory and will provide important theoretical support for improving the prediction accuracy of structural fatigue life analysis methods.

I am a student of Mr. Gao, and probably, I am the student who has received the most personal guidance from Mr. Gao in recent years. Whether it was midsummer or hot summer, I sat in his small study room of less than $10\,m^2$ for hours. When he learned that I was teaching Mechanics of Materials for undergraduates, he not only passed me the classical textbook and lecture notes but also carefully prepared the lecture materials, discussed with me many times, and told me how to use these materials to enhance the effect of knowledge transfer and creative ability cultivation in the classroom. He emphasized the importance of taking a "big gradient, small steps", "identify and solve students' problems using their perspective and language", etc. Mr. Gao not only innovated in scientific research but also always spearheaded the development of teaching and education ideas and concepts. For 70 years, the Beihang community have taken it as their responsibility to solve major basic scientific problems and key core technologies in China's aerospace field and to cultivate high-quality talents in the aerospace field as their mission. For 70 years, Mr. Gao has led many disciples, colleagues, and students, embodying the original mission and responsibility of the Beihang people through his actions.

I was asked by Mr. Gao to write a Foreword for *Intelligent Fatigue Statistics*, which he co-authored with Mr. Xu Jiajin, and I am really humbled to do so. "Look up and it's endless, drill down and it's solid". I hope to do my best to inherit the mantle of Mr. Gao's commitment to the development of China's independent theory of aircraft structure life determination and keep moving forward to meet his expectations.

Rui Bao
Beijing, China

Preface

THE CHINESE VERSION OF *Intelligent Fatigue Statistics (IFS)*, co-authored by Academician Zhentong Gao and Mr. Jiajin Xu, was published by Beihang University Press in Beijing in October 2022 (Gao & Xu, 2022). This book represents significant modifications based on Mr. Gao Zhentong's *Fatigue Applied Statistics* (Gao, 1986) and *Fatigue Reliability* (Gao & Xiong, 2000). It incorporates the latest achievements in intelligent research on fatigue statistics, denoted by Xu (2021), Xu & Gao (2022), and Xu (2022). The major modifications include introducing the Weibull distribution (WD) as a full-state distribution and safe life (one of the three parameters of the Weibull distribution). This book extensively explores and discusses computer intelligence, particularly in the context of the three-parameter Weibull distribution, resulting in some preliminary outcomes such as the Zhentong Gao method and the Generalized Zhentong Gao method (Xu, 2023). These methodologies simplify the calculation of the three parameters of the Weibull distribution (WD) without relying on the cumbersome derivation and complex calculations of traditional methods.

The English version of *Intelligent Fatigue Statistics* has made necessary revisions to the Chinese version and added research results after the Chinese version was published, such as the addition of the Generalized Zhentong Gao method for estimating the three parameters of WD (Xu, 2023). It was also emphasized that taking the logarithm of fatigue data for further research and analysis is inappropriate, as it would distort the physical meaning of fatigue life (Xu & Gao, 2023).

This book can serve as a reference book for seniors and graduate students majoring in aviation, machinery, architecture, and related engineering in higher education institutions. It is also a valuable reference for engineers and researchers working in the fields of fatigue, reliability engineering, and so on.

This book is structured into three main parts.

The first part is "Mathematical Foundations of Intelligent Fatigue Statistics". This part emphasizes two crucial new concepts of WD: (1) WD is a full-state distribution (including left-patrial, right-patrial, symmetric, and exponential distributions); and (2) the safe life (i.e., location parameter, one of the three parameters of WD) represents the physical meaning of 100% reliability. The significance of safe life is also the theoretical basis of the Zhentong Gao method (Xu, 2021). At the same time, it is pointed out that people often take logarithms of asymmetric fatigue life data to make the data appear more symmetrical, so that Gaussian distribution can be used to fit. But, this processing method changes the physical meaning of the data itself, which is inappropriate. WD should be used to directly fit the original data (Xu & Gao, 2023).

The second part is "Computer Fundamentals of Intelligent Fatigue Statistics". This part encourages readers to be proficient in using Excel, especially Python. Python is highlighted as superior to C++ and Java, as it can entirely replace Basic and Fortran, as well as statistical computing software packages such as MATLAB®, SPSS, SAS, and R. In a sense, Python is indispensable to IFS.

The third part is "Some Applications of Intelligent Fatigue Statistical". This section is the most crucial part of the book. It uses the foundational knowledge from the first two parts to solve some typical examples in "fatigue statistics", allowing readers to understand how to address practical problems. The problems and data presented are derived from real-world scenarios. A new algorithm for estimating the three parameters of the Weibull distribution, namely, the Zhentong Gao method, is introduced. This method not only intelligently addresses the determination of the three parameters of the Weibull distribution but also effectively deals with the fitting of three-parameter fatigue performance data. Building on the Zhentong Gao method, this book further provides confidence intervals for the three parameters of the Weibull distribution and fatigue life. Further combining with MLE, the Generalized Zhentong Gao method is proposed, which avoids the tedious derivation and complicated calculation of MLE in determining the three parameters of the Weibull distribution (Xu, 2023).

The final chapter explicitly clarifies that machine learning is not mysterious; in a sense, machine learning is essentially a form of statistical learning, and digital experiments are one of its important methods. Following this perspective, this book presents some digital experiments conducted in fatigue statistics and statistical learning, as well as a representative Bootstrap (Xu, 2022). It is hoped that these will assist readers in achieving advanced stages of "intelligent fatigue statistics".

We would like to thank all those who have been helpful in publishing this book, especially Dr. Sun Lian from CRC Press and her team (including outstanding representative Ms. Feng Xiaoyin). Without their effective cooperation, this book could not have been published so quickly. We want to thank Professor Bao Rui from Beihang University for his irreplaceable work in the publication of this book; thank you to Mr. Wan Weihao for his unconditional support; we want to thank Dr. Yang Mingqing (Grant M Yang) and Ms. Elayna Ferguson for their careful proofreading; otherwise, the quality of this book could not be guaranteed. We want to thank Mr. Ni Wanmei, Dr. Nabil Chbouki, Ms. Sonia Zamir, and Dr. Wang Dajian who did an excellent polishing job on this book. Of course, we also would like to thank our respective families for their heartfelt contributions.

REFERENCES

Gao ZT (1986), *Fatigue Applied Statistics*, National Defense Industry Press, Beijing.

Gao ZT & Xiong JJ (2000), *Fatigue Reliability*, Beihang University Press, Beijing.

Gao ZT & Xu JJ (2022), *Intelligent Fatigue Statistics*, Beihang University Press, Beijing.

Xu JJ (2021), The Gao Zhentong Method in the Intelligentization of Fatigue Statistics, *Journal of Beijing University of Aeronautics and Astronautics*, 47(10): 2024–2033. doi: 10.13700/j.bh.1001-5965.2020.0373

Xu JJ (2022), Digital Experiment for Estimating Three Parameters and Their Confidence Intervals of Weibull Distribution, *International Journal of Science, Technology and Society*, 10(2): 72–81. doi: 0.11648 /j.ijsts. 20221002.16

Xu JJ & Gao ZT (2022), Further Research on Fatigue Statistics Intelligence, *Acta Aeronautical et Astronautical Sinica*, 43(8). doi: 10.7527/S1000. 6893.2021.25138

Xu JJ (2023), Generalized Zhentong Gao Method for Estimating Three Parameters of Weibull Distribution, *SCIREA Journal of Computer*, 8(2). doi: 10.54647/ computer520345

Xu JJ & Gao ZT (2023), From Gaussian Distribution to Weibull Distribution, *Global Journal of Researches in Engineering: I Numerical Methods*, 23(1): 1–6. doi: 10.34257/GJREIVOL23IS1PG1

I

Mathematical Foundations of Intelligent Fatigue Statistics

CHAPTER 1

Basic Knowledge of Probability Theory

1.1 PROBABILITY, RANDOM EVENT, AND RANDOM VARIABLE

If everything in the world were "determined", there would be no such thing as random events. Random events are inseparable from "uncertainty". So, is the world absolutely determined or completely uncertain? In reality, it is neither entirely certain nor entirely uncertain. For example, celestial movements are generally determined, allowing for relatively accurate predictions of astronomical phenomena such as sunrise, sunset, and eclipses. However, this is not unconditional and is subject to time constraints. On the contrary, for uncertain events like rolling dice, people may not know the outcome of each throw, but the probability of a particular number appearing is definite, provided the assumption that the dice are very uniform holds true. Probability theory is a branch of mathematics that studies the laws governing the occurrence of random (uncertain) events and holds significant practical importance. Mathematical statistics, built on the foundation of probability theory, is an applied science that helps identify relevant patterns when the essential relationship between certain properties is temporarily unknown. For instance, in engineering, where the microscopic quantification of fatigue life is still unavailable, mathematical statistics can help discover correlated factors causing component damage due to fatigue, thereby preventing unnecessary accidents.

This is particularly crucial for aircraft, where knowledge about when to repair or replace components after a certain number of flights is essential. Currently, there is only empirical data, making mathematical analysis of this data crucial. Random variables are needed to describe these random events. However, before rigorously defining random events and random variables, several basic concepts related to probability need to be defined. Their importance lies in serving as the foundation of probability theory, aiding in understanding various so-called random phenomena.

1.1.1 A Few Basic Concepts about Probability

1.1.1.1 Random Experiment

An experiment with an uncertain outcome, satisfying the following three conditions (Meyer, 1986):

a. The experiment can be repeated under essentially constant conditions.

b. The experiment has more than one possible outcome, all of which can be known in advance.

c. Each experiment results in only one outcome, but the outcome cannot be known in advance.

Example 1.1

Taking the classic example of rolling a die, the outcome of the number appearing on the die is uncertain. It can be verified that this satisfies the three conditions of a random experiment: the experiment of rolling a die can be repeated, all possible outcomes of the experiment are known (1, 2, 3, 4, 5, 6), and the result of each trial can only be one of these numbers, but which one cannot be predicted in advance.

1.1.1.2 Sample Space

The set of all possible outcomes of a random experiment.

Example 1.2

Using the example of rolling a die whose sample space is {1, 2, 3, 4, 5, 6}.

Example 1.3

The sample space for flipping a coin is {H, T}. And the sample space for flipping two coins is {HH, HT, TH, TT}.

1.1.1.3 Probability

Let E be a random experiment, so that S is associated with a sample space, for each event A in S corresponds to a real number, and satisfy the following three conditions is called the probability of A, denoted as P(A). The following three conditions (sometimes also known as probability axioms) are (Fisz, 1978):

a. $0 \leq P(A) \leq 1$ (the probability must be a positive real number not greater than 1);

b. If A and B are mutually exclusive and their intersection is the empty set, i.e., $P(A \cap B) = 0$, then $P(A \cup B) = P(A) + P(B)$ (axiom of addition);

c. $P(S) = 1$ (normalization principle).

Example 1.4

If $\emptyset$ is the empty set, then $P(\emptyset) = 0$, assuming A be non-empty. Therefore, A is mutually exclusive with $\emptyset$ and $P(A) = P(A \cup \emptyset) = P(A) + P(\emptyset)$, i.e. $P(\emptyset) = 0$.

Example 1.5

If $\overline{A}$ is the complement of A, then $P(\overline{A}) = 1 - P(A)$. In fact, $S = \overline{A} \cup A$ and $P(S) = 1$, so, $P(\overline{A}) = 1 - P(A)$

Example 1.6

If A and B are not mutually exclusive, we have:

$$P(A \cup B) = P(A) + P(B) - P(A \cap B)$$

Not mutually exclusive means that their intersection is not the empty set, as shown in Figure 1.1. The meaning of the above equation is very clear. Why should it subtract $P(A \cap B)$? Because of double counting. Therefore, if two events are mutually exclusive, it means that $A \cap B = \emptyset$. Therefore, the addition axiom is still true.

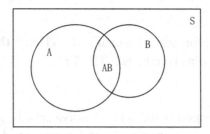

FIGURE 1.1   Venn diagram.

1.1.1.4 Conditional Probability

The fatigue life of an aircraft depends not only on its flight time but also on its takeoff and landing frequency. This implies that the probability of the aircraft's fatigue life is different under different conditions. In mathematical terms, the conditional probability is defined as:

$$P(A|B) = P(A \cap B)/P(B) = P(AB)/P(B) \tag{1.1}$$

where A and B are two events in the same sample space S. $P(A|B)$ is the probability of event A under the condition where event B occurs. In the future, for the convenience of not causing misunderstandings, $A \cap B$ will be written as AB. As shown in Figure 1.1,

$$P(B|A) = P(AB)/P(A) \tag{1.2}$$

From equations (1.1 and 1.2), we can derive:

$$P(AB) = P(A)P(B|A) = P(B)P(A|B) \tag{1.3}$$

1.1.1.5 Mutual Independence

Conditional probability emphasizes the correlation of two events, while mutual independence emphasizes the irrelevance of two events, i.e., the occurrence of event A is not related to event B and vice versa. This means that:

$$P(B|A) = P(B); \ P(A|B) = P(A) \rightarrow P(AB) = P(A)P(B) \tag{1.4}$$

This is a necessary and sufficient condition for two events to be mutually independent. It's essential to distinguish independence from mutual exclusivity discussed earlier. Two mutually exclusive events mean: $P(AB) = 0$, indicating two mutually exclusive events cannot occur at the same time. In contrast, two independent events generally have $P(AB) \neq 0$, to emphasizing only $P(AB) = P(A)P(B)$.

Example 1.7

Taking the example of flipping a coin twice, assuming that the first coin toss comes up heads as event A, and the second time tails occur is event B. Then, it is clear that the two events are independent of each other, i.e., $P(AB) = P(A) \times P(B)$. But they are not mutually exclusive, i.e., $P(AB) \neq 0$. Further, suppose that event AB is regarded as event C, then events C and A are neither mutually exclusive nor independent of each other, because $P(CA) = P(C) \neq P(C)P(A)$. Generally speaking, under certain conditions, if the probability of the sum of two events can be directly added, they are mutually exclusive, while if their product can be directly multiplied, they are independent of each other.

Example 1.8

Assuming the probability of damage to the tires and brackets of the landing gear of a certain type of aircraft is p_1 and p_2, respectively, what is the reliability (probability) of the landing gear "assembly" in use? It is obvious that these two components are independent of each other, so the reliability is: $P(UT) = P(U) P(T) = (1 - p_1)(1 - p_2)$, where U and T represent that the tires and brackets of the landing gear work properly, respectively, while $P(U)$ and $P(T)$ represent the probability of their proper operation, respectively.

1.1.1.6 Bayes Formula

From equation (1.3), it can be obtained that:

$$P(A|B) = P(B|A)P(A)/P(B) \tag{1.5}$$

The Bayesian formula and the generalized Bayes' Theorem are very important and will not be expanded here; the interested reader can refer to any textbook on probability statistics. Only one example is given here for reference (Fisz, 1978).

Example 1.9

Consider two cannons with hit rates of 0.8 and 0.7, firing 9 and 10 rounds, respectively. What is the probability that the second cannon hits the target given that only one cannon hits it?

Solution

In fact, it can be assumed that A_1 and A_2 are the firing events of guns 1 and 2, respectively, while B indicates hitting the target. Then, according to the question, it is obvious that according to the Bayesian formula we can get:

$$P(A_2|B) = P(A_2)P(B|A_2)/[P(A_1)P(B|A_1) + P(A_2)P(B|A_2)]$$

where

$$P(A_1) = 9/19, \; P(A_2) = 10/19, \; P(B|A_1) = 0.8, \; P(B|A_2) = 0.7$$

Therefore,

$$P(A_2|B) = [(10/19) \times 0.7]/[(9/19) \times 0.8 + (10/19) \times 0.7] = 0.493.$$

This result shows there is really not much difference in the probability of the two guns hitting the target under these conditions.

1.1.2 Random Event

Any subset in the sample space of a random experiment is a random event. For example, if a dice is rolled and the sample space is {1,2,3,4,5,6}, then the occurrence of the number 3 is a random event of the dice roll, and of course, the occurrence of "1 or 2" is also a random event. However, the occurrence of "1 and 2" is not a random event because it is not a subset of the sample space of a single dice thrown. Instead, it can be a subset of two dice rolled at the same time. That is to say, the so-called random event is related to the sample space of the random experiment, and it is meaningless to talk about any random event if the random experiment and its sample space are not clearly stated.

1.1.3 Random Variable

What is the relationship between random variable and random events? In a sense, a random variable (Fisz, 1978) can be thought of as a single-valued real function of random event. Note that a random variable is essentially a function, not an independent variable in the usual sense. For example, if a coin is flipping, the sample space is {heads, tails}, the random events are "heads" and "tails", and the random variable X(heads)=1 and X(tails)=0. At this point, the random variable quantifies the random events quantified, which is conducive to mathematical operations. In the future, random variable is usually expressed in upper-case Latin letters or lowercase Greek letters. An important role of random variable is that it expresses the probability of a random event easily, still taking the coin flipping as an example, under normal circumstances, the probability of both heads and tails is 1/2, which can be conveniently written as, $p(X=1)=1/2$, $p(X=0)=1/2$.

There are two types of random variables in terms of type, discrete and continuous. The examples given above, such as flipping a coin or rolling a dice, are discrete. However, the random variables in the fatigue problem are continuous. For example, the random variable of fatigue life for a random experiment on a group of specimens at the same stress level is continuous. In the real world, the mathematical sense of continuous does not exist. It can only be an infinite approximation of rational numbers to real numbers, just like real numbers can only achieve a certain degree of accuracy in computers. The so-called continuity is also relative.

1.2 DISTRIBUTION, PROBABILITY DENSITY FUNCTION, AND RELIABILITY OF RANDOM VARIABLE

1.2.1 Distribution Function and Probability Density Function of a Random Variable

One of the most important functions of a random variable as independent variable is the distribution function and its associated density function. The so-called distribution function is used to describe the probability distribution of a random variable, such as the fatigue life of the aircraft. People are not interested in a point in time probability, but the probability of exceeding this point in time. Thus, the distribution function of random variable is defined as:

$$F(x) = P(X < x) \tag{1.6}$$

Here, it should be noted that the capital letter X indicates a random variable, while the lowercase letter x indicates a real value it may take. Sometimes, the distribution function $F(x)$ can also be referred to as the cumulative distribution function (CDF). This definition for continuous or discrete random variable is valid, but the strict definition also requires knowledge of the real variable function, which is ignored here. The most important feature of CDF is:

$$F(-\infty) = 0, \ F(\infty) = 1 \tag{1.7}$$

The sufficient conditions for a single-valued real variable function to be a distribution function are (1) non-decreasing and at least left semi-continuous, and (2) satisfying condition (1.7). For a discrete random variable, one can have:

$$F(x) = \Sigma_{x_i < x} p_i \tag{1.8}$$

For example, rolling dice: $F(1.5) = 1/6$, while $F(3.1) = 1/2$, $F(6.2) = 1$, etc.

For a continuous random variable, one has:

$$F(x) = \int_{-\infty}^{x} f(x)dx \tag{1.9}$$

where f(x) is called the probability density function (PDF) of the random variable X, henceforth referred to as density or PDF. By definition, we have:

$$F(\infty) = \int_{-\infty}^{\infty} f(x)dx = 1 \tag{1.10}$$

From the calculus point of view, we immediately get:

$$f(x) = F'(x) \tag{1.10a}$$

The probability for a continuous random variable may not necessarily be an improbable event even if it is zero (however, it should be noted that the probability of an impossible event must be zero), and similarly, the probability is 1 that may not be a necessary event (but the probability of an inevitable event is always 1). For example, X is a continuous function with density f(x), and a is any point, according to the definition of the probability of this point 'a', it can be found as:

$$P(X = a) = \int_{a}^{a} f(x)dx = 0$$

Setting that $R' = \{R - a\}$, which means the set of real numbers except for the point 'a', so that $P(R') = \int_{R} f(x)dx = 1$.

It can be seen that the probability of a continuous random variable at any point is zero, so it is also meaningless. Therefore, for a continuous random variable, we can only say what is the probability of being in a certain interval (a, b):

$$P(a < x < b) = \int_a^b f(x)dx \qquad (1.11)$$

1.2.2 Reliability and Destruction Rate (Gao & Xiong, 2000)

For fatigue statistics, N is generally used to represent the fatigue life as a random variable, but making the theory discussed general, it is better to use x instead of N as a random variable. This means that f(x) is generally used to represent the PDF of some distribution. This way, for any specified x_p, the probability of being less than x_p can be determined. This means that from the point of view of fatigue life, the probability that the life is less than 5.4 is 5%, which can be defined as destruction rate (Gao & Xiong, 2000). This gives the physical meaning of the distribution function $F(x_p)$. And the probability of greater than 5% is called reliability level (hereafter referred to as reliability, as shown in Figure 1.2), so it is not difficult to see that the so-called reliability is essentially a probability. This may be one reason why reliability is represented by P instead of R (Gao & Xiong, 2000):

$$P = 1 - F(x_p) \qquad (1.12)$$

and destruction rate,

$$D = 1 - P = F(x_p) \qquad (1.13)$$

Therefore,

$$P + D = 1 \qquad (1.14)$$

These relationships are very important, and this formula is often used later. You can look back at Example 1.8.

1.3 EXPECTATION AND MOMENTS OF A RANDOM VARIABLE

1.3.1 Expectation of a Random Variable and Arithmetic Mean in Statistics

Characterizing a random variable with numbers is crucial. One of the most important numbers is the mathematical expectation or mean (Fisz,

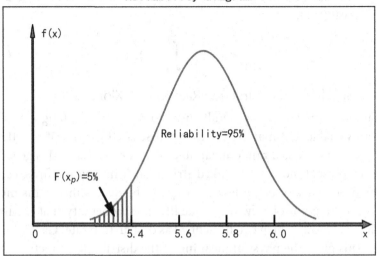

FIGURE 1.2   Reliability diagram.

1978). For a discrete random variable g(X), the mathematical expectation is defined as:

If

$$\Sigma_{k=1}^{\infty} p_k \left| g(x_k) \right| < \infty \tag{1.15}$$

then:

$$\Sigma_{k=1}^{\infty} p_k g(x_k) \tag{1.16}$$

For continuous random variables with density f(x), the Expectation is defined as:

If

$$\int_{-\infty}^{\infty} \left| g(x) \right| f(x) dx < \infty \tag{1.17}$$

then:

$$\int_{-\infty}^{\infty} g(x) f(x) dx \tag{1.18}$$

which can be written as $E[g(X)]$. It is also important to emphasize that the conditions (1.15) and (1.17) are essential, that is, to make equations (1.16) and (1.18) converge. If equations (1.15) and (1.17) do not hold, then the so-called mathematical expectation is also meaningless, because there may be "infinite" possibilities. It is worth noting that the expectation of the random variable X is exactly the "center of gravity" (center of form, center of mass) of the area enclosed by PDF, so this is the geometric meaning of expectation. This is because:

$$\int_{-\infty}^{\infty} xf(x)dx \Big/ \int_{-\infty}^{\infty} f(x)dx = E(X)$$

Another important concept is the "median" x_{50} determined by the following equation:

$$\int_{-\infty}^{x_{50}} f(x)dx = 1/2$$

It is obvious that if f(x) is symmetric, then $\mu = x_{50}$.

By definition, it is obvious that if the random variable X as well as c are constants, then we have:

$$E(X+c) = E(X) + c \tag{1.19}$$

and,

$$E(cX) = cE(X) \tag{1.20}$$

Special attention should be paid here where the distinction between the two concepts of expectation (sometimes called mean) in probability theory and the arithmetic mean (average) in statistics is different. The former is a parameter associated with a theoretical probability distribution, while the arithmetic mean is a statistical concept that relates to the specific outcome of each randomized trial. Of course, the two are closely related, as the "Law of Large Numbers" (Fisz, 1978) points out when samples tend to infinity, the arithmetic mean of samples tends to the mathematical expectation of population.

1.3.2 The Significance of the Moments of Each Order of Random Variable

As we all know, mathematical expectations are indeed very important and widely used, but it is not enough to simply use mathematical expectations to describe random variables. The so-called "dispersion degree" still needs to be introduced:

$$D(X)=Var(X)=\sigma^2=E[(X-E(X))^2] \tag{1.21}$$

is mean square error (MSE) or Variance, and its square root σ is standard deviation (SD). The reason for taking the square root is to make its dimension (unit) consistent with the expectation. This number can better reflect the "degree of dispersion" of the random variable.

By using the concept of "moment" in mechanics, we can introduce the concept of so-called moments of each order of the random variable, which is also intended to describe some characteristics of the random variable.

If $g(X)=X^k$, then its mathematical expectation is m_k, which is called the k-rank moment of the random variable X. In particular, for $k=1$, it is the mathematical expectation of X, therefore:

$$m_1=E(X) \tag{1.22}$$

Let's look again at the relationship between the mean squared deviation and the first- and second-order moments. In fact:

$$Var(X)=E[(X-E(X))^2]=E[X^2-2XE(X)+E^2(X)]$$

Notice that $E(X)$ is a constant, i.e., $E[2XE(X)]=2E(X)*E(X)$, and $E(X^2)=m_2$ so that the above formula can become,

$$Var(X)=m_2-m_1^2 \tag{1.23}$$

This shows that the mean squared deviation of a random variable is its second-order moments minus the square of its first-order moments. That is, if the first-order moment (mathematical expectation) is zero, then the mean squared difference is the second-order moment. No wonder people like to call a random variable with a mathematical expectation of zero and a second-order moment of one a "standard random variable", i.e.:

$$E(X)=0, Var(X)=\sigma^2=1 \tag{1.24}$$

The non-standard random variable can be turned into standard random variable by linear transformation, which will greatly facilitate the study of non-standard random variable. For example, let Y be a non-standard random variable and X be a standard random variable, we have:

$$X = (Y - E(Y))/\sigma \qquad (1.25)$$

And,

$$E(X) = E(Y - E(Y)) = E(Y) - E(Y) = 0 \qquad (1.26)$$

On the other hand:

$$Var(X) = D[(Y - E(Y))/\sigma] = VAR(Y)/\sigma^2 = 1 \qquad (1.27)$$

Note here that $VAR(X + c) = VAR(X)$, $E[(aX)^k] = a^k E(X^k)$

Here, we would like to supplement the concept of the so-called center moment, which in physics is to take the moment of the center of gravity, and here is to take the moment of the mathematical expectation (i.e., the center of gravity of the probability). The SD mentioned earlier is the second-order center moment, and in general, we have:

$$\mu_1 = E(X - E(X)) = 0, \quad \mu_2 = E\left[(X - E(X))^2\right] = \sigma^2 = m_2 - m_1^2;$$

$$\mu_3 = E\left[(X - E(X))^3\right] = m_3 - 3m_1 m_2 + 2m_1^3 \qquad (1.28)$$

We can define:

$$C_s = \mu_3/\sigma^3 \qquad (1.29)$$

is coefficient of skewness to measure the degree of asymmetry in data distribution (Gao & Xiong, 2000).

REFERENCES

Gao ZT & Xiong JJ (2000), *Fatigue Reliability*, Beihang University Press, Beijing, p. 42, 50.

Fisz M (1978), *Wahrscheinlichkitsrechnung und Mathematische Stastistik*, Shanghai Science and Technology Press, Shanghai, pp. 11–13, 22–24, 27, 61, 163.

Meyer PL (1986), *Introduction Probability and Statistics*, Higher Education Press, Beijing, p. 9.

CHAPTER 2

Random Variable Functions and Moment Generating Functions

2.1 RANDOM VARIABLE FUNCTION AND TWO-DIMENSIONAL RANDOM VARIABLE

2.1.1 Meaning of Random Variable Function

In Section 1.1.3, it was mentioned that a random variable itself is also a type of function. However, what's more important is that a random variable can also serve as an independent variable, thus forming a function of random variables. For example, when randomly selecting n samples from a population to measure their fatigue life $(x_1, ..., x_n)$, their average value is, $\tilde{X} = (1/n) \sum_{i=1}^{n} x_i$. And this fatigue life is a random variable, which can be represented by X. Therefore, this sample average $\tilde{X}$ is a random variable function of the random variable X which is obviously also a random variable. It is also obvious that the statistical parameters of the sample (e.g., mean and variance) are all random variable functions, i.e., they are all random variables. However, the parameters of the population are "constant", although they are often "unknown" and need to be estimated using the parameters of the sample. This is precisely the meaning of statistics.

DOI: 10.1201/9781003488477-3

For a random variable function $Y = H(X)$, the main concern is how to conveniently obtain the PDF $g(x)$ of Y based on the PDF $f(x)$ of the random variable X.

Example 2.1

Suppose: $f(x) = \begin{cases} \dfrac{1}{2}, -1 < x < 1 \\ 0, \text{other cases} \end{cases}$ and $Y = H(X) = X^2$ (Meyer, 1986).

Let the CDF and PDF of Y be $G(y)$ and $g(y)$, respectively. Therefore, by definition, we have:

$$G(y) = P(Y \leq y) = P(X^2 \leq y) = P(-y^{1/2} \leq X \leq y^{1/2}) = F(y^{1/2}) - F(-y^{1/2})$$

and,

$$g(y) = G'(y) = f(y^{1/2})/(2y^{1/2}) + f(-y^{1/2})/(2y^{1/2}) = [f(y^{1/2}) + f(-y^{1/2})]/(2y^{1/2}) \quad (2.1)$$

It is worth pointing out that this example, although it gives the specific form of $f(x)$, actually "doesn't work". That is, (2.1) is independent of the specific form of $f(x)$.

2.1.2 Two-Dimensional Random Variable

The random variables mentioned above are all one-dimensional. However, many random variables are functions of two or more random variables, which can be called a two-dimensional random variable or a multi-dimensional random variable. In fact, similar to the case of multivariate function, as long as the two-dimensional case is clearly studied, then it is not difficult to extend to the multi-dimensional case. For example, there are two groups of test pieces presented by the two different processing methods, which are implied by the random variables X and Y, respectively. The fatigue lives of these two groups are considered as the X and Y variables. If you want to compare the two groups for their advantages and disadvantages, their difference $Z = X - Y$ needs to be calculated. And this difference is a two-dimensional random variable. Similar to the distribution and PDF of one-dimensional random variable, two-dimensional random variable distribution and PDF (Gao, 1986) are given as:

$$P(x < X < x + dx, y < Y < y + dy) = p(x, y)dxdy \quad (2.2)$$

The probability densities of three-dimensional and even multi-dimensional random variable can be given similarly.

According to the nature of PDF, it must be normalized, which requires:

$$\int_{-\infty}^{\infty}\int_{-\infty}^{\infty} p(x,y)dxdy = 1 \qquad (2.3)$$

It can be set:

$$f(x)= \int_{-\infty}^{\infty} p(x,y)dy; \quad g(y)= \int_{-\infty}^{\infty} p(x,y)dx \qquad (2.4)$$

If the random variables X and Y are independent of each other, then:

$$p(x, y)=f(x)g(y) \qquad (2.5)$$

In general, we have:

$$p(x, y)=p(x)p(y|x)=p(y)p(x|y) \qquad (2.6)$$

Here, p(x, y) is generally called the joint PDF of x, y. And, f(x) and g(y) are called the marginal PDF of x and y, respectively.

Example 2.2

Study from 1 to 21 sets of numbers; the random variables X and Y are taken from this set of parity (even for 1, odd for 0) and can be divided by 3 (can be divided by 3 otherwise 0) properties (Fisz, 1978) (Table 2.1):

This table shows that the marginal probabilities with respect to X and Y are, respectively,

$$P(X=1)=10/21, \; P(X=0)=11/21 \text{ and } P(Y=1)=7/21=1/3,$$
$$P(Y=0)=2/3.$$

TABLE 2.1   Distribution Characteristics of Numbers from 1 to 21

	X=1	X=0	(Y)Total
Y=1	3	4	7
Y=0	7	7	14
(X)Total	10	11	21

2.2 MATHEMATICAL EXPECTATION AND VARIANCE OF THE SUM OF RANDOM VARIABLES

2.2.1 Mathematical Expectation of the Sum of Random Variables

Let's consider random variables X and Y with probability density functions f(x) and g(y), respectively. Also, let Z be a two-dimensional random variable defined as $Z = X + Y$, with corresponding joint probability density function p(x, y). What is the mathematical expectation of Z? We have:

$$E(Z) = E(X+Y) = \int_{-\infty}^{\infty} \int_{-\infty}^{\infty} (x+y) p(x,y) dxdy$$

$$= \int_{-\infty}^{\infty} \int_{-\infty}^{\infty} xp(x,y) dxdy + \int_{-\infty}^{\infty} \int_{-\infty}^{\infty} yp(x,y) dxdy$$

Noting equation (2.4), it can be obtained:

$$E(Z) = \int_{-\infty}^{\infty} xf(x)dx + \int_{-\infty}^{\infty} yg(y)dy = E(X) + E(Y) \qquad (2.7)$$

This shows that the above formula holds regardless of whether the random variables X and Y are independent or not.

It is also not difficult to generalize to the case of the sum of n random variables:

$$E(X_1, X_2, ..., X_n) = E(X_1) + E(X_2) + ... + E(X_n) \qquad (2.8)$$

As mathematical expectation for the difference between two random variables, we have:

$$E(X - Y) = E(X) - E(Y) \qquad (2.9)$$

2.2.2 Variance of the Sum of Random Variables

Let's further investigate the variance of $Z = X + Y$. Based on the definition of variance:

$$Var(Z) = Var(X+Y) = \int_{-\infty}^{\infty} \int_{-\infty}^{\infty} [(x+y) - E(X+Y)]^2 p(x,y)dxdy \qquad (2.10)$$

Noting that (2.7), we have:

$$Var(Z) = \int_{-\infty}^{\infty} \int_{-\infty}^{\infty} \left[(x - E(X)) + (y - E(Y)) \right]^2 p(x, y) dxdy$$

$$= \int_{-\infty}^{\infty} \int_{-\infty}^{\infty} \left[x - E(X) \right]^2 p(x, y) dxdy$$

$$+ \int_{-\infty}^{\infty} \int_{-\infty}^{\infty} \left[y - E(Y) \right]^2 p(x, y) dxdy \qquad (2.11)$$

$$+ 2 \int_{-\infty}^{\infty} \int_{-\infty}^{\infty} (x - E(X))(E(Y) - y) p(x,y) dxdy$$

It is easy to see:

$$\int_{-\infty}^{\infty} \int_{-\infty}^{\infty} \left[x - E(X) \right]^2 p(x, y) dxdy = \int_{-\infty}^{\infty} \left[x - E(X) \right]^2 f(x) dx = Var(X)$$

Similarly, we have:

$$\int_{-\infty}^{\infty} \int_{-\infty}^{\infty} \left[y - E(Y) \right]^2 p(x, y) dxdy = Var(Y)$$

Let's set:

$$Cov(X, Y) = \int_{-\infty}^{\infty} \int_{-\infty}^{\infty} (x - E(X))(y - E(Y)) p(x,y) dxdy \qquad (2.12)$$

and Cov(X, Y) is called the covariance of the random variables X and Y.

$$Var(X + Y) = Var(X) + Var(Y) + 2Cov(X, Y)$$

Clearly, if X and Y are independent of each other, then:

$$Cov(X, Y) = 0 \qquad (2.13)$$

$$Var(X + Y) = Var(X) + Var(Y) \qquad (2.14)$$

And for Var(X − Y), the same can be obtained that:

$$Var(X-Y) = Var(X) + Var(Y) - 2Cov(X, Y) \qquad (2.15)$$

However, it should be noted that when X and Y are independent of each other, they are not subtracting variances but still adding together:

$$Var(X-Y) = Var(X) + Var(Y) \qquad (2.16)$$

As for the sum (or difference) of the n mutually independent random variable is still the sum of the individual variances:

$$Var(X_1, X_2, \ldots, X_n) = Var(X_1) + Var(X_2) + \ldots + Var(X_n) \qquad (2.17)$$

Regarding the covariance, let's add a point, which will be useful in the future. The correlation coefficient of the random variable X and Y is defined by the following formula:

$$r = Cov(X,Y)/[Var(X)Var(T)]^{1/2} = Cov(X,Y)/\sigma_X\sigma_Y \qquad (2.18)$$

$r=0 \leftrightarrow Cov(X,Y) \rightarrow X$ and Y are independent of each other.

$r>0 \leftrightarrow Cov(X,Y) \rightarrow$ both are positively correlated otherwise negatively correlated.

2.3 MOMENT GENERATING FUNCTIONS AND THEIR PROPERTIES

2.3.1 Definition of Moment Generating Functions

The concept of generating functions, although unfamiliar to many, was introduced more than 200 years ago by the renowned French mathematician Laplace and later developed by Euler. A generating function (Meyer, 1986) is essentially a power series, and its coefficients correspond to a certain sequence as if this sequence were generated by this "generating function". So, what is the use of this generating function? It is often used in combinatorics.[1] However, here we mainly discuss its role in calculating moments of random variables, referred to as the "moment generating function", which greatly simplifies the computation of moments of any order. Let's introduce the moment generating function $M_\xi(\theta)$ of a random variable ξ:

$$M_\xi(\theta) = E(exp(\theta\xi)) \qquad (2.19)$$

and,

$$M_\xi(\theta) = \int_{-\infty}^{\infty} f(x)\exp(\theta x)dx \qquad (2.20)$$

We can expand $\exp(\theta x)$ according to a Taylor series. Equation (2.20) can be become:

$$M_\xi(\theta) = \int_{-\infty}^{\infty} f(x)\left(1+\theta x+(\theta x)^2/2!+(\theta x)^3/3!+\ldots\right)dx \qquad (2.21)$$

so:

$$\left[dM_\xi(\theta)/d\theta\right]_{\theta=0} = \int_{-\infty}^{\infty} f(x)xdx = m_1 \qquad (2.22)$$

and,

$$\left[d^k M_\xi(\theta)/d\theta^k\right]_{\theta=0} = \int_{-\infty}^{\infty} f(x)x^k dx = m_k; \ k=2,3.,4,\ldots \qquad (2.23)$$

From this, we can see that having these equations makes it relatively easy to calculate moments of any order of the PDF, which is very convenient for computing related parameters of the PDF.

2.3.2 Properties of Moment Generating Functions (Gao & Xiong, 2000)

There are three important properties (theorems) of moment generating functions. Using the definition of the moment generating function, it is easy to derive:

Theorem 2.1

If the random variable x and y have the following linear relationship, $Y = aX+b$, then:

$$M_\eta(\theta) = \exp(b\theta)M_\xi(a\theta) \qquad (2.24)$$

It can be easily seen that if

$$a=-1, b=0 \rightarrow M_{-X}(\theta)=M_X(-\theta) \qquad (2.25)$$

Theorem 2.2

If the random variables X and Y are independent of each other, then:

$$M_{X+Y}(\theta)=M_X(\theta)M_Y(\theta) \qquad (2.26)$$

It is not difficult to prove this theorem. In fact, X and Y are independent of each other, i.e., $f(x,y)=f(x)g(y)$. Therefore:

$$M_{X+Y}(\theta)= \int_{-\infty}^{\infty} \exp\left[\theta(x+y)\right]f(x)g(y)dxdy = M_X(\theta)M_Y(\theta)$$

It is not difficult to generalize to the case of X-Y. Of course, X and Y are required to be independent of each other. That is:

$$M_{X-Y}(\theta)=M_X(\theta)M_Y(-\theta) \qquad (2.27)$$

For n independent random variable, you can set, $Y=a_1X_1+a_2X_2+...+a_nX_n$, we have that:

$$M_Y(\theta)=M_{X1}(a_1\theta)M_{X2}(a_2\theta)...M_{Xn}(a_n\theta) \qquad (2.28)$$

where

$$M_{Xi}(a_i\theta)=E\left[\exp(a_i\theta x_i)\right]= \int_{-\infty}^{\infty} \exp(a_i\theta x_i)f(x)dx, \ i=1,2,...,n \qquad (2.29)$$

Theorem 2.3

The probability distribution of a random variable is uniquely determined by its moment generating function.

This theorem is also known as the "uniqueness theorem" and its significance is evident, particularly for Gaussian distributions. The proof for general distributions is somewhat involved, and interested readers can refer to the relevant textbooks (Fisz, 1978).

NOTE

1 Refer to https://blog.csdn.net/weixin _ 40730615/article/details/102834969

REFERENCES

Fisz M (1978), *Wahrscheinlichkitsrechnung und Mathematische Stastistik*, Shanghai Science and Technology Press, Shanghai, p. 45.

Gao ZT (1986), *Fatigue Applied Statistics*, National Defense Industry Press, Beijing, p. 105.

Gao ZT & Xiong JJ (2000), *Fatigue Reliability*, Beihang University Press, Beijing, p. 175.

Meyer PL (1986), *Introduction Probability and Statistics*, Higher Education Press, Beijing, p. 115, 267.

CHAPTER 3

Several Commonly Used Distributions

3.1 GAUSSIAN DISTRIBUTION

Most probability theory textbooks introduce several commonly used discrete distributions, such as the binomial distribution, Poisson distribution, geometric distribution, and so on. However, in the field of fatigue statistics, most distributions used are continuous. Therefore, we will not discuss these discrete distributions here. Instead, we will focus on continuous distributions, including the Gaussian distribution (GD), the gamma distribution, the beta distribution, and particularly, the Weibull distribution (WD).

3.1.1 The Characteristics of Gaussian Distribution

The Gaussian distribution (often referred to as the Normal Distribution) is characterized by the PDF with the following form for a random variable X:

$$f(x) = [1/(2\pi)^{1/2}\sigma]\exp[-(x - \mu)^2/2\sigma^2] \tag{3.1}$$

also known as Gaussian distribution. Notice that:

$$E(X) = \left[1/(2\pi)^{1/2}\sigma\right] \int_{-\infty}^{\infty} x \exp\left[-(x-\mu)^2/2\sigma^2\right]dx$$

$$= \left[1/(2\pi)^{1/2}\sigma\right] \int_{-\infty}^{\infty} (t+\mu) \exp[-t^2/2\sigma^2]dt = \mu \tag{3.2}$$

DOI: 10.1201/9781003488477-4

Here, we used the result of the probability integral:

$$\int_{-\infty}^{\infty} \exp(-u^2) du = \pi^{1/2} \tag{3.3}$$

And,

$$E(X^2) = \left[1/(2\pi)^{1/2} \sigma\right] \int_{-\infty}^{\infty} x^2 \exp[-(x-\mu)^2/2\sigma^2] dx$$

$$= \left[1/(2\pi)^{1/2} \sigma\right] \int_{-\infty}^{\infty} (t+\mu)^2 \exp\left[-t^2/2\sigma^2\right] dt = \sigma^2 + \mu^2 \tag{3.4}$$

Because it can be set $v = t/2^{1/2} \sigma$, we have:

$$\int_{-\infty}^{\infty} t^2 \exp(-t^2/2\sigma^2) dt = -2^{1/2} \sigma^3 \exp(-v^2)|_{-\infty}^{\infty} + 2^{1/2} \sigma^3 \int_{-\infty}^{\infty} \exp(-v^2) dv$$

$$= (2\pi)^{1/2} \sigma^3$$

So that,

$$Var(X) = E(X^2) - [E(X)]^2 = (\sigma^2 + \mu^2) - \mu^2 = \sigma^2$$

This shows that the mathematical expectation of the normal distribution is μ and the variance is σ^2. For convenience later, N (μ, σ^2) can be used to denote a normal distribution with mathematical expectation μ and variance σ^2 and its standard form is:

$$[1/(2\pi)^{1/2}] \exp(-x^2/2) \tag{3.5}$$

where the mean $\mu = 0$ and the deviation $\sigma^2 = 1$ is the **standard normal distribution** (SND).

The Gaussian distribution has the following characteristics based on its definition (Gao, 1986):

1. It is unimodal (single-peaked).

2. It is symmetric, with the mode (peak), median, and mean all being the same.

3. Its curve resembles a bell shape, often referred to as a "bell-shaped" distribution.

4. It is ubiquitous, as a significant portion of random variables encountered in real-life situations closely or approximately follows the Gaussian distribution. Even in arbitrary distributions, the distribution of the mean will tend to approximate a Gaussian distribution for large samples.

5. It is simple, as it only requires two parameters (μ, σ^2) to determine the entire distribution's shape.

Due to these favorable characteristics, the Gaussian distribution is the most widely studied and applied distribution. However, it is evident that not all data follow a Gaussian distribution, and in most cases, data following a Gaussian distribution is only an approximate description. Subsequent chapters will demonstrate that data on fatigue life often do not precisely follow Gaussian distribution but are better described by the Weibull distribution. Sometimes, taking the logarithm of fatigue life data makes it more closely resemble a Gaussian distribution, but this is still an approximation. Therefore, we need to introduce the Weibull distribution and study it in more depth.

3.1.2 Standard Fraction

In practical applications of the Gaussian distribution, people are familiar with probabilities associated with deviations of ±1, ±2, ±3 standard deviations from the mean, which correspond to probabilities of 68.26%, 95.5%, and 99.74%, respectively. However, real-world situations often involve deviations that are not precisely equal to integer multiples of standard deviations. Therefore, the concept of standard scores is introduced to describe how far a value is from the mean in terms of standard deviations:

$$z = \text{Standard Fraction} = (x - \mu)/\sigma \tag{3.6}$$

Here, x represents the actual value. This concept is crucial in practical applications.

Example 3.1: Reliability and Destruction
Rate of Gaussian Distribution

The concepts of reliability and destruction rate were introduced in Section 1.2.2 and are used here only in GD. It is clear that:

$$P = \left[1/(2\pi)^{1/2} \sigma \right] \int_{x_p}^{\infty} \exp\left[-(x-\mu)^2 / 2\sigma^2 \right] dx \tag{3.7}$$

This is the reliability. And destruction rate can be expressed as:

$$D = F(x_p) = \left[1/(2\pi)^{1/2}\sigma\right] \int_{-\infty}^{x_p} \exp\left[-(x-\mu)^2/2\sigma^2\right]dx \quad (3.8)$$

Therefore: $P + D = 1$; this is exactly equation (1.14).

For GD x_p actually corresponds to the "standard fraction" u_p of Gaussian distribution:

$$u_p = (x_p - \mu)/\sigma \quad (3.9)$$

With this formula, it is very easy to understand the meaning and calculation of x_p in GD. Further, we can set:

$$u = (x - \mu)/\sigma \quad (3.10)$$

From equation (3.7), we have

$$P = \left[1/(2\pi)^{1/2}\sigma\right] \int_{x_p}^{\infty} \exp\left[-(x-\mu)^2/2\sigma^2\right]dx$$

$$\quad (3.11)$$

$$= \left[1/(2\pi)^{1/2}\right] \int_{u_p}^{\infty} \exp\left(-u^2/2\right)du$$

This shows that P can be expressed not only by the area enclosed by the normal PDF, $f(x) = [1/(2\pi)^{1/2}\sigma] \exp[-(x-\mu)^2/2\sigma^2]$. It can also be expressed by the area enclosed by the standard normal PDF: $\varphi(u) = [1/(2\pi)^{1/2}] \exp(-u^2/2)$.

The advantage of using the standard fraction is that the corresponding "standard" value can be obtained by computer. For example, the reliability value of P = 0.99 is defined. Note that the standard fraction table is given as:

$$D = F(x_p) = 1 - P \quad (3.12)$$

The standard score corresponding to 0.01 obtained on Excel is that, u_p = NORMSINV (0.01) = −2.32635. It is almost the same as −2.326 in Table 5-1 of (Gao, 1986). As for the reliability to be improved to 0.999 → 0.001, the corresponding standard fraction u_p = −3.09023 can also be obtained through Excel, which is almost identical to Table 5-1 in (Gao,1986). If we take the reliability as 0.5, it can be obtained that:

$$u_p = 0, \text{ and } x_{50} = \mu \quad (3.13)$$

Example 3.2

It is known that the population logarithmic mean life of a material $\mu = 5.8$ and the population standard deviation $\sigma = 0.14$; try to find the safe life with a reliability of 98%.

Solution

Since the reliability is 0.98, we can find a table that 0.02 corresponds to $u_{98} = -2.056$.

Using equation (3.9), we can find: $x_p = \mu + \sigma \times u_p$ and $x_{98} = 5.8 - 0.14 \times 2.056 = 5.512$

So, $\log N_{98} = 5.512$ and $N_{98} = 10^{5.512} = 3.25 \times 10^5$ (times).

Clearly, if the standard deviation doubles:

$x_{98} = 5.8 - 0.28 \times 2.056 = 5.224$; therefore, $N_{98} = 10^{5.224} = 1.67 \times 10^5$ (times)

This indicates that the life has almost halved, implying that a larger standard deviation means the material's quality is less stable.

It's worth noting that, for Gaussian distribution, there is no 100% reliability. Additionally, with the help of Excel, there is no need to look up tables, and you can directly obtain values corresponding to standard scores. This issue will be detailed in Chapter 5 of this book.

3.1.3 Probability Density Function of the Sum and Difference of Normal Variables

Assuming X is a Gaussian-distributed random variable, as shown in Section 2.3, we can derive its moment generating function:

$$M_X(\theta) = \left[1/(2\pi)^{1/2} \sigma \right] \int_{-\infty}^{\infty} \exp(\theta x) \exp\left[-(x-\mu)^2 / (2\sigma^2) \right] dx \quad (3.14)$$

By making the variable substitution: $z = (x - \mu)/(2^{1/2}\sigma)$, we can get: $x = (2^{1/2}\sigma)z + \mu$, $dx = (2^{1/2}\sigma)dz$.

Therefore:

$$M_X(\theta) = \left(1/\pi^{1/2} \right) \int_{-\infty}^{\infty} \exp\left[-\left(z - 2^{-1/2}\sigma\theta \right)^2 \right] \exp\left(\theta\mu + \theta^2\sigma^2/2 \right) dz$$

and,

$$M_X(\theta) = \exp(\theta\mu + \theta^2\sigma^2/2) \quad (3.15)$$

Now, assuming that X and Y are two mutually independent normal variables, we have:

$$M_X(\theta) = \exp\left(\theta\mu_1 + \theta^2\sigma_1^2/2\right); \quad M_Y(\theta) = \exp\left(\theta\mu_2 + \theta^2\sigma_2^2/2\right)$$

By Theorem 2.2 from Section 2.3, we find: $M_{X+Y}(\theta) = M_X(\theta)M_Y(\theta)$; therefore:

$$M_{X+Y}(\theta) = \exp\left[\theta(\mu_1 + \mu_2) + \theta^2\left(\sigma_1^2 + \sigma_2^2\right)/2\right] \tag{3.16}$$

Clearly, this expression is similar in form to equation (3.15), indicating that the distribution of X + Y is also Gaussian, with mean $\mu = \mu_1 + \mu_2$ and variance $\sigma^2 = \sigma_1^2 + \sigma_2^2$. Therefore, the PDF of X + Y is:

$$f(x) = \left[1 \middle/ \left(2\pi\left(\sigma_1^2 + \sigma_2^2\right)\right)^{1/2}\right]\exp\left[-\left(x - (\mu_1 + \mu_2)\right)^2 \middle/ \left(2\left(\sigma_1^2 + \sigma_2^2\right)\right)\right] \tag{3.17}$$

Similarly, for the PDF of the difference of two independent Gaussian-distributed variables X and Y, we have:

$$f(x) = \left[1 \middle/ \left(2\pi\left(\sigma_1^2 + \sigma_2^2\right)\right)^{1/2}\right]\exp\left[-\left(x - (\mu_1 - \mu_2)\right)^2 \middle/ \left(2\left(\sigma_1^2 + \sigma_2^2\right)\right)\right] \tag{3.18}$$

Note the plus and minus signs for both here: the mathematical expectations are subtracted, but the variances are still added.

3.1.4 "Issues" with Gaussian Distribution

In fact, there are no inherent "issues" with the Gaussian distribution itself. The problem arises from people's strong preference for the simplicity of the Gaussian distribution. Gaussian distribution is well-suited for symmetric data, but it falls short when dealing with asymmetric data. To make structural fatigue life data fit the Gaussian distribution more closely, people often take the logarithm of the data to make it appear symmetric. However, this distorts the physical interpretation of the structural fatigue life data, especially with regard to the critical parameter of safety life (Xu & Gao, 2023). Additionally, it is important to note that Gaussian distribution theoretically allows for negative lifetimes, which is naturally unacceptable. Therefore, the Weibull distribution, which is better suited for structural fatigue life data, needs to be introduced.

3.2 WEIBULL DISTRIBUTION

3.2.1 Origin and Characteristics of Weibull Distribution

As mentioned in Section 3.1.4, the Gaussian distribution is well-suited for fitting symmetric data but struggles with asymmetric data. The Weibull distribution, on the other hand, excels at fitting asymmetric data and plays a particularly important role in the field of fatigue reliability. The Weibull distribution was developed by Waloddi Weibull (1887–1979), a Swedish engineer, scientist, and mathematician. According to online sources, "In 1939, he published a paper on the Weibull distribution in probability theory and statistics". In 1951, "he submitted his most famous paper on Weibull distribution to the American Society of Mechanical Engineers (ASME) through seven case studies".[1] In 1961, he published a book on Weibull distribution for "fatigue analysis", which is likely a significant reason for introducing the Weibull distribution in fatigue studies (Weibull, 1961). Here's a brief introduction to the advantages and disadvantages of the Weibull distribution compared to the Gaussian distribution:

1. Theoretically, the safe life at 100% reliability can be provided (Xu, 2021).

2. WD is a full-state distribution (Xu, 2021), which is more suitable for fitting asymmetric data, but even symmetric data is sometimes indistinguishable from Gaussian distributions. In this sense, GD can be regarded as a "first-order approximation" of WD. That is to say, WD represents a more general distribution.

3. While the Weibull distribution has many advantages in "fatigue statistics", it has a more complex mathematical form, and parameter estimation is more challenging. This complexity is a major reason why it is less widely applied (McCool, 2012).

3.2.2 Probability Density Function of Weibull Distribution

The PDF of the Weibull distribution is given by Gao (1986):

$$f(N) = [b/(N_a - N)][(N - N_0)/(N_a - N_0)]^{b-1} \exp\{-[(N - N_0)/(N_a - N_0)]^b\} \quad (3.19)$$

or

$$f(x) = [b/(x_a - x_0)][(x - x_0)/(x_a - x_0)]^{b-1} \exp\{-[(x - x_0)/(x_a - x_0)]^b\} \quad (3.20)$$

where N_0 is the (reliability of 100%) safe life or in other fields x_0 is called the location parameter, while N_a is the characteristic life or x_a which is called the characteristic parameter, b is the shape parameter; and $N_0 < N < \infty$.

Or from another point of view, one can set:

$$\lambda = (x_a - x_0) \tag{3.21}$$

therefore,

$$f(x) = (b/\lambda)[(x - x_0)/\lambda)]^{b-1}\exp\{-[(x - x_0)/\lambda]^b\} \tag{3.22}$$

λ is called the scale or scaling parameter. In terms of importance, it may be a little more appropriate to replace the feature parameter by the scale parameter λ. Therefore, (3.22) is hereafter used instead of (3.19) as the general Three-Parameter Weibull Distribution expression.

If $x_0 = 0$, (3.22) becomes Two-Parameter Weibull Distribution density function:

$$f(x) = (b/\lambda)(x/\lambda)^{b-1}\exp[-(x/\lambda)^b] \tag{3.23}$$

It should be noted here that sometimes people often replace b and λ^{-b} in the above formula with α and λ, respectively, in order to be better looking formally, so that the two-parameter Weibull distribution is written as follows:

$$f(x) = \alpha\lambda x^{\alpha-1}\exp(-\lambda x^\alpha) \tag{3.23a}$$

However, the physical meaning of λ is not obvious at this point, so the form of (3.23a) is not used in this book.

If we set $\lambda = 1$, it becomes the One-Parameter Weibull Distribution density function:

$$f(x) = bx^{b-1}\exp(-x^b) \tag{3.23b}$$

In Figure 3.1, the PDF of Weibull distribution with the same location parameter $x_0 = 0.5$ but different shape parameter b and different scale parameter λ is given.

From the PDF of Weibull distribution and its graph, we can further observe the following characteristics:

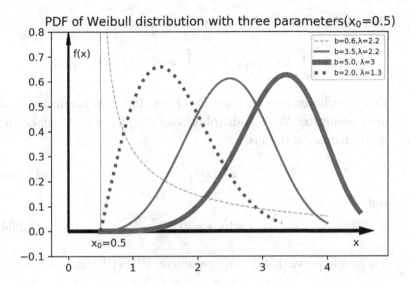

FIGURE 3.1   Weibull Distribution (where $x_0 = 0.5$) PDF.

1. The shape parameter b of the Weibull distribution is of great significance. When $0 < b < 1$, it resembles a $1/x$ function; when $b = 1$, it becomes the exponent distribution. For $1 < b < 3$, it represents a left-partial distribution, and for $3 < b < 4$, it approximates the Gaussian distribution. When $b > 4$, it becomes right-partial. Furthermore, when $b = 2$, it corresponds to the Rayleigh distribution.[2]

2. The Weibull density function has three parameters, one more than the Gaussian distribution. This is both an advantage (for describing safety life) and a challenge (due to its more complex mathematical form). There are pros and cons, but overall, the advantages outweigh the disadvantages. It can describe symmetric as well as asymmetric data, making it a "full-state distribution".

The above description provides a qualitative overview of the Weibull distribution. Now, let's delve into a quantitative analysis. To facilitate this, we will initially conduct a detailed study of the one-parameter Weibull distribution, and similar results can be obtained for the two-parameter and three-parameter Weibull distributions:

1. The one-parameter WD's PDF satisfies the normalization condition, which means:

$$\int_0^\infty f(x)dx = 1 \qquad (3.24)$$

This can be proven by defining $z = x^b$, leading to

$$\int_0^\infty f(x) = \int_0^\infty bx^{b-1}\exp(-x^b) = \int_0^\infty \exp(-z)dz = 1 \rightarrow q.e.d$$

This conclusion holds true for both the two-parameter and three-parameter Weibull distributions. We just need to make the appropriate substitutions:

$$z = (x/\lambda)^b \tag{3.25a}$$

and,

$$z = [(x - x_0)/\lambda]^b \tag{3.25b}$$

2. It is easy to prove that the one-parameter WD's CDF is:

$$F(x_p) = 1 - \exp(-x_p^b) \tag{3.26}$$

In fact, the same as (1) can be set

$$z = x^b \rightarrow F(x_p) = \int_0^{x_p} f(x)dx = 1 - \exp(-x_p^b) \rightarrow q.e.d.$$

And this distribution function is precisely the destruction rate, and the reliability P is then, that:

$$P = 1 - F(x_p) = 1 - D = \exp(-x_p^b) \tag{3.27}$$

This equation is very important because it conveniently gives the functional form of the reliability, which is of great significance for applications. While the two- and three-parameter Weibull distributions are, respectively:

$$P = 1 - F(x_p/\lambda) = \exp[-(x_p/\lambda)^b] \text{ and } P = \exp\{-[(x_p - x_0)/\lambda]^b\} \tag{3.27a}$$

3. The density function is upper convex when $b > 1$, so there must be an extreme (large) value point. In fact:

$$f'(x) = b\exp(-x^b)[(b-1)x^{b-2} - bx^{b-1}x^{b-1})] = b\exp(-x^b)x^{b-2}(b - 1 - bx^b)$$

Setting $f'(x) = 0$ lead to $b - 1 - bx^b = 0$, which yields:

$$x^b = (b-1)/b \tag{3.28}$$

Therefore,

$$x_{max} = \exp\{\ln[(b-1)/b]/b\} \tag{3.29}$$

Then, find the second-order derivative of f(x),

$$f''(x) = b \exp(-x^b)x^{b-3}\{[(b-2) - bx^b](b-1-bx^b) - b^2x^b\}$$

Notice equation (3.28), we have that:

$$f''(x_{max}) = b \exp(-x^b)x^{b-3}b(1-b) < 0$$

This proves that x_{max} is indeed an extreme value of f(x) and that f(x) is an upper convex function. And noting (3.29) yields that:

As b gets larger, $\ln[(b-1)/b] \to 0$, $x_{max} \to 1$. At this point, the peak is concentrated on 1.

For the two- and three-parameter Weibull PDF, the location of the maximum point is, respectively,

$$x_{max} = \exp\{\ln[(b-1)\lambda^b/b]/b\}, \tag{3.29a}$$

and,

$$x_{max} = x_0 + \exp\{\ln[(b-1)\lambda^b/b]/b\} \tag{3.29b}$$

4. The mathematical expectation (mean) of the WD can be derived as follows:

$$E(X) = \int_0^\infty xbx^{b-1}\exp(-x^b)dx = \int_0^\infty z^{(1+1/b)-1}\exp(-z)dz$$

Using the definition of the gamma function,

$$E(X) = \Gamma(1+1/b) \tag{3.30}$$

Similarly, for the two- and three-parameter Weibull distributions, the mathematical expectation is, respectively:

$$E(X) = \lambda\Gamma(1+1/b) \text{ and } E(X) = x_0 + \lambda\Gamma(1+1/b) \tag{3.30a}$$

5. The variance of the Weibull distribution can also be calculated:

$$Var(X) = E(X^2) - E^2(X) \tag{3.31}$$

With $E(X^2) = \int_0^\infty x^2 b x^{b-1} \exp(-x^b) dx = \int_0^\infty z^{(1+2/b)-1} \exp(-z) dz$, we get:

$$E(X^2) = \Gamma(1+2/b) \qquad (3.32)$$

This leads to the variance formula:

$$Var(X) = \Gamma(1+2/b) - \Gamma^2(1+1/b) \qquad (3.33)$$

And for both two- and three-parameter Weibull distributions, the variance is:

$$Var(X) = \lambda^2[\Gamma(1+2/b) - \Gamma^2(1+1/b)] \qquad (3.33a)$$

It is clear that the variance of the two- and three-parameter WD is not directly related to x_0.

6. A discussion on the similarity of WD and GD.
   As mentioned earlier, when $3 < b < 4$, the two distributions are similar. What does this similarity imply? The Gaussian distribution signifies symmetry, with the mean, median, and mode all coinciding. The Weibull distribution does not possess this property. However, under certain conditions, it is possible to make two of these values coincide. For instance, setting the mean and median to be equal is one such condition. The mean can be obtained from equation (3.31), and the median can be found when $p = 50\%$ in equation (3.27): $0.5 = \exp(-x_{50}b)$. In this case, x_{50} can be replaced with the mean from equation (3.30), leading to the transcendental equation:

$$\ln(2) = \Gamma(1+1/b)^b \qquad (3.34)$$

This is a transcendental equation, which can be solved using Excel, but it's a bit tricky. However, utilizing Python gives the result quickly. But either method utilizes Newton's dichotomy. For this purpose, the above equation can be rewritten as:

$$g(b) = \Gamma(1 + 1/b)^b - \ln(2) = 0 \qquad (3.35)$$

Using Excel, the following results can be obtained:

If it is accurate to 10^{-4}, then this b = 3.44 is acceptable (Table 3.1).

And the result of running in Python, whose code will be presented in Section 6.3, is that

k(number of pairs of fractions) = 19, E(precision) = 1e-08, b = 3.43954

The advantage of Excel is that it is intuitive, but it requires human intervention in the calculation process and is less efficient.

Incidentally, this conclusion, although introduced from the one-parameter Weibull distribution, is applicable to both two- and three-parameter distributions, since the symmetry is only related to the shape parameter b.

Also, according to equation (1.28), the skewness coefficient of Weibull distribution is that:

$$C_s = [\Gamma(1 + 3/b) - 3\Gamma(1 + 1/b)\Gamma(1 + 2/b)$$
$$+ 2\Gamma^3(1 + 1/b)]/[\Gamma(1 + 1/b) - \Gamma^2(1 + 1/b)]^{3/2} \qquad (3.36)$$

Therefore, if the bias coefficient is zero, it should also be the time when the symmetry is best. From this point of view, let the above equation be zero, i.e.:

$$\Gamma(1 + 3/b) - 3\Gamma(1 + 1/b)\Gamma(1 + 2/b) + 2\Gamma^3(1 + 1/b) = 0 \qquad (3.37)$$

This is another transcendental equation with respect to b. Again, the following result can be obtained using the code introduced in Section 6.3, that:

k(number of pairs of points) = 26, E(accuracy) = 1e-08, b = 3.60235

This result shows it is close to b = 3.44 but does not overlap, which is to be expected since Weibull distribution cannot be perfectly symmetrical after all.

TABLE 3.1   Used Excel to Solve the Equation

b	2	4	3	3.5	3.45	3.43	3.44
g(b)	9.23E-02	−2.00E-02	1.89E-02	0.00E+00	−3.92E-04	3.60E-04	−1.73E-05

3.3 GAMMA AND BETA DISTRIBUTIONS

3.3.1 Gamma and Beta Functions

The gamma function (Gao & Xiong, 2000) has been used in the previous talk about Weibull distribution and is only briefly introduced here:

1. **Definition**: The gamma function is a generalized integral

$$\Gamma(\alpha) = \int_0^\infty x^{\alpha-1} \exp(-x) dx \qquad (3.38)$$

where $\alpha > 0$. The result of the above integral is 1 when $\alpha = 1$, i.e., $\Gamma(1) = 1$.

2. **Properties**: One of the most important properties is that

$$\Gamma(\alpha+1) = \alpha\Gamma(\alpha) \qquad (3.39)$$

For positive integer values of α, this simplifies to $\Gamma(n+1) = n!$, where

$$\Gamma(1) = 0! = 1 \qquad (3.40)$$

If we make a variable substitution on the right side of (3.38), such as $x = y^2$, we get:

$$\Gamma(\alpha) = 2\int_0^\infty y^{2\alpha-1} \exp(-y^2) dy \qquad (3.41)$$

Closely related to the gamma function is the beta function.

1. **Definition**: The beta function (Gao & Xiong, 2000) is also a generalized integral:

$$B(m, n) = \int_0^1 x^{m-1} (1-x)^{n-1} dx \qquad (3.42)$$

where, m, n > 0

2. **Properties**: It is obvious that it is symmetric, meaning $B(m, n) = B(n, m)$ If we make the substitution: $x = \sin^2\theta$, (3.42), it becomes:

$$B(m, n) = 2\int_0^{\pi/2} \sin^{2m-1}\theta\cos^{2n-1}\theta d\theta \qquad (3.43)$$

Also, using (3.41), we can show that:

$$\Gamma(m)\Gamma(n) = 4\int_0^\infty x^{2m-1}\exp(-x^2)dx \int_0^\infty y^{2n-1}\exp(-y^2)dy \quad (3.44)$$

Using polar coordinates here, this leads to:

$$\Gamma(m)\Gamma(n) = B(m, n)\Gamma(m + n) \quad (3.45)$$

which can be rearranged as:

$$B(m, n) = \Gamma(m)\Gamma(n)/\Gamma(m + n) \quad (3.46)$$

When both m and n are positive integers, we have:

$$B(m, n) = [(m - 1)! \, (n - 1)!]/(m+n - 1)! \quad (3.47)$$

3.3.2 Gamma Distribution

Gamma distribution is also a distribution that will be used later, and its PDF is:

$$f(x) = [\beta^\alpha/\Gamma(\alpha)]x^{\alpha-1}\exp(-\beta x) \quad (3.48)$$

where $\alpha, \beta > 0, 0 < x < \infty$. α and β are called shape and scale parameters, respectively. Again, its mathematical expectation and variance can be calculated. By definition, we have:

$$E(X) = \int_0^\infty \left[\beta^\alpha/\Gamma(\alpha)\right]x^* x^{\alpha-1}\exp(-\beta x)\,dx \quad (3.49)$$

For the purpose of integration, let $y = \beta x$, the above equation can become:

$$[1/\beta\Gamma(\alpha)]\int_0^\infty y^\alpha\exp(-y)dy = [1/\beta\Gamma(\alpha)]\Gamma(\alpha+1)$$

which leads to variance:

$$E(X) = \alpha/\beta \quad (3.50)$$

And for the variance:

$$\sigma^2 = \mathrm{Var}(X) = E(X^2) - [E(X)]^2 \quad (3.51)$$

Noting that: $E(X^2) = \int_0^\infty [\beta^\alpha/\Gamma(\alpha)]x^2 * x^{\alpha-1}\exp(-\beta x)dx$, and doing the variable substitution $y = \beta x$,

by similar calculations as before, you can find:

$$E(X^2) = \int_0^\infty [1/\beta^2\Gamma(\alpha)]y^{\alpha+1}\exp(-y)dy = \Gamma(\alpha+2)/[\beta^2\Gamma(\alpha)] = \alpha(\alpha+1)/\beta^2$$

(3.52)

Therefore:

$$\sigma^2 = Var(X) = \alpha(\alpha+1)/\beta^2 - (\alpha/\beta)^2 = \alpha/\beta^2 \qquad (3.53)$$

3.3.3 Beta Distribution

Beta distribution is also a distribution that will be used later, such as the rank distribution to be introduced in the next chapter is a beta distribution. The PDF is:

$$f(x) = [\Gamma(\alpha+\beta)/\Gamma(\alpha)\Gamma(\beta)]x^{\alpha-1}(1-x)^{\beta-1} \qquad (3.54)$$

It can be shown that: $\int_0^1 f(x)dx = 1$. Similarly, you can calculate that (Mao et al., 2006):

$$E(X^k) = \alpha(\alpha+1)...(\alpha+k-1)/[(\alpha+\beta)(\alpha+\beta+1)...(\alpha+\beta+k-1)]$$

$$= \Gamma(\alpha+k)\Gamma(\alpha+\beta)/[\Gamma(\alpha)\Gamma(\alpha+\beta+k)] \qquad (3.55)$$

which leads to:

$$E(X) = \alpha/(\alpha+\beta); \quad Var(X) = \alpha\beta/[(\alpha+\beta)^2(\alpha+\beta+1)] \qquad (3.56)$$

NOTES

1 Refer to https://en.wikipedia.org/wiki/Waloddi_Weibull
2 A density function with a shape such as $f(x) = (x/\sigma^2)\exp(-x^2/\sigma^2)$ is called the Rayleigh density function.

REFERENCES

Gao ZT (1986), *Fatigue Applied Statistics*, National Defense Industry Press, Beijing, p. 57, 70.

Gao ZT & Xiong JJ (2000), *Fatigue Reliability*, Beihang University Press, Beijing, p. 87, 89.

McCool JI (2012), *Using the Weibull Distribution*, John Wiley & Sons, Inc., Hoboken, New Jersey.

Mao SS, Wang JL & Pu XL (2006), *Advanced Mathematical Statistics*, 2nd, Higher Education Press, Beijing, pp. 9–10.

Weibull W (1961), *Fatigue Testing and Analysis of Results*, Pergamon Press, Oxford.

Xu JJ (2021), Zhentong Gao method in the fatigue statistics intelligence, *Journal of Beijing University of Aeronautics and Astronautics*, 47(10): 2024–2033. doi: 10.13700/j.bh.1001-5965.2020.0373

Xu JJ & Gao ZT (2023), From Gaussian distribution to weibull distribution, *Global Journal of Researches in Engineering: I Numerical Methods*, 23(1). doi: 10.34257/GJREIVOL23IS1PG1

CHAPTER 4

Basic Knowledge of Statistics

4.1 MEANING OF STATISTICS

The meaning of "statistics" in Chinese is "summary calculation", while the English word "statistics" is derived from "the Latin word Status, which means the state and condition of various phenomena. From this root comes the Italian word Stato, which means the concept of 'state' and also contains the meaning of the structure of the state and knowledge of the country".[1]Thus, statistics appeared much earlier than probability. Still, the emergence of statistics as a modern science was not until the seventeenth century, while modern statistics had to wait until the twentieth century.

In the current era of artificial intelligence (AI), modern statistics has become indispensable and has shown its prowess. Statistics has rapidly developed in the process of analyzing and utilizing big data with the help of computers, creating various new algorithms based on statistics. In some sense, it can be said that understanding statistics is essential in the AI era. Of course, for those involved in fatigue-related research and applications, statistics is an indispensable tool.

The father of modern statistics, British statistician Carl Pearson (1857–1936), believed that "the lack of quantitative study of biological phenomena is not acceptable, and was determined to make the theory of evolution further quantitative description and analysis on the basis of a general qualitative account. He continued to use statistical methods to make

DOI: 10.1201/9781003488477-5

new contributions to biology, genetics, and eugenics. At the same time, he imported many new concepts based on the probabilistic studies of his predecessors who were good at gambling on opportunities, distilled bio-statistical methods into general methods for handling statistical information, developed statistical methodology, and fused both probability theory and statistics into one furnace". [2] For example, he introduced the concept of standard deviation, and he gave the famous χ^2 (chi-square) test; in short, he contributed a lot and still influences statistics today. Below is a simple example of a statistical application.

Example 4.1: Estimating the Number of Fish in a Lake

Suppose there is only one type of fish in a lake and its population is N. You capture u fish from the lake, mark them, and then release them back into the lake. After some time, you capture V fish from the lake and find that x of them have marks. How can you estimate the total number of fish, N?

This is a practical statistical application problem. It's apparent that marked fish are uniformly distributed in the lake, so we can write:

$$u/N = x/V \rightarrow N = uV/x.$$

In statistical terms, this means that the density of fish in the sample is the same as the density in the entire lake. If there are multiple fish species in the lake, the method is similar. Of course, this estimation should rely on a reasonably sized sample; it should not be too small. However, determining the appropriate sample size is not arbitrary but will be discussed quantitatively in later chapters.

Thus, statistics has two main meanings: the "original meaning" of quantifying various political and economic data and statistical inference, which involves inferring the population distribution and its parameters from sample data. Estimating the fatigue life of components from a sample of fatigue life data is one of the problems this book will address.

4.2 STATISTICS AND PROBABILITY AND RELATED LAWS

4.2.1 Statistics and Probability

In Chapter 1, the theoretical (classical) definition of probability was introduced, which is based on the sample space. For example, when rolling a dice, the sample space is {1, 2, 3, 4, 5, 6}, and under normal circumstances, each number is equally likely to occur:

$$P(1) = P(2) = P(3) = P(4) = P(5) = P(6) = 1/6 \tag{4.1}$$

There is also an empirical (frequency) definition of probability, which states: "Under certain conditions, repeat an experiment n times, where n_A is the number of times event A occurs. If, as n increases, the frequency n_A/n stabilizes around a value p, then p is called the probability of event A under those conditions, denoted as $P(A) = p$. This definition is called the statistical definition of probability".[3] For example, if you really roll the dice and record the number of points that occur, when n is small, equation (4.1) is definitely not valid, and as n increases, equation (4.1) becomes the "limiting case". This is the result of what is known as the law of large numbers.

A famous experiment that demonstrates this principle is: "In 1939, South African mathematician Krech carelessly ran to Europe and ended up in a concentration camp. He found an interesting pastime for himself: he flipped a coin 10,000 times and recorded the number of times it landed heads up (the probability). As the number of coin flips increased, the probability of heads clearly approached 50%".[4]

4.2.2 Law of Large Numbers

The law of large numbers is based on the foundation of frequency probability. "In mathematics and statistics, the Law of Large Number describes the result of performing a large number of repetitions of the same experiment. According to this law, as the sample size increases, the average tends to approach the expected value".[5] A more rigorous proof of this law can be found in Appendix I or in textbooks on probability theory and statistics.

To deepen the understanding of the law of large numbers, let's consider a common misconception known as the "gambler's fallacy" (Bennett et al., 2016).

Imagine a game of flipping a (fair) coin, where getting heads wins you $1 and getting tails loses you $1. Suppose you've already flipped the coin 100 times, and you've seen 45 heads and 55 tails. People often believe that according to the law of large numbers, the probabilities of heads and tails should approach 1/2, so they should start "winning" soon. However, in reality, it doesn't happen, and they continue to lose more money. Has the law of large numbers failed? The fact is that the law of large numbers is still valid; it's just being misinterpreted. Look at Table 4.1 to understand why (Bennet, 2016).

TABLE 4.1 Experimental Results of Flipping Coin

Rolling times	Head times	Prob. of Head	Dif. of Head and Tail times
100	45	45%	10
1,000	470	47%	60
10,000	4,950	49.50%	100
1,00,000	49,900	49.90%	200

The table clearly shows that as the number of flipping coin increases, the probability of a head increases and gets closer to 0.5, but the absolute difference between heads and tails also increases, i.e., you lose more and more. The gambler's fallacy is not a problem with law of large numbers, but rather a lack of understanding of the "independent event" that each coin toss is an unrelated event and will not necessarily result in heads after a few tails in a row. For example, if two girls are born in a row, will the next one be a boy? Will the probability increase? Generally speaking, if there is no manual intervention, then the probability of having a boy is still 1/2 because each birth is an independent event.

4.3 POPULATION AND SAMPLE

4.3.1 Basic Terms in Statistics

In general statistical applications, the term "population" refers to the entire set of objects under study. The population size is usually large, and it is neither practical nor necessary to survey every individual in the population. Statistical inference aims to solve this problem by estimating the characteristics of the population using a sample. To ensure accuracy, it is essential to define the terms commonly used:

n: Size of the Sample, i.e., the number of samples taken at a time, not to be confused with the number (m) of samples taken.

$\bar{X}$: Sample Mean (Average), which is the statistical mean derived from actual sampled numbers.

μ: Population Mean (Average), which is a theoretical value and the expected value of a random variable.

σ: Population Standard Deviation, which is also a theoretical value defined in Section 1.3.2.

S^2: Sample Variance, defined as:

$$S^2 = (1/n)\sum_{i=1}^{n}\left(x_i - \tilde{X}\right)^2 \tag{4.2}$$

s: Sample Standard Deviation. It is important to note that it is defined as:

$$s^2 = \left[1/(n-1)\right]\sum_{i=1}^{n}\left(x_i - \tilde{X}\right)^2 \tag{4.3}$$

Here, n is the number of samples taken and $\tilde{X}$ is the mean value of that sample. But why the sample standard variance should be divided by $n-1$, and this is also the origin of the so-called "unbiased estimation". Here is a brief explanation of what is meant by "unbiased estimation". Suppose a parameter w* of the sample is an estimate of the population parameter w, and E(w*)=w, then this w* is called an "unbiased estimate" of w; otherwise, it is a "biased estimate". Obviously, according to the law of large numbers, $E(\tilde{X})=\mu$, i.e., the sample mean $\tilde{X}$ is an unbiased estimate of the population mean. Similarly, it can be shown that $E(S^2)\neq\sigma^2$. Therefore, S^2 is a biased estimate of σ^2 and $E(s^2)=\sigma^2$, so s^2 is an unbiased estimate of σ^2. However, since the mathematical derivation is more complicated, the interested reader can refer to Appendix II.

Sometimes, $n-1$ is deliberately called as "degrees of freedom" to distinguish between n and $n-1$, which probably refers the concept of "degrees of freedom" from the mechanics.[6] Actually, the meaning of both is similar. In statistics, "degrees of freedom" is the number of independent or freely varying data in a sample when statistics of the sample is used to estimate the parameters of the total. Thus, "when estimating the population mean (μ), the Degree of Freedom is n since all n numbers in the sample are independent of each other and any number not yet drawn is not affected by any value already drawn". However, "the statistic used in estimating the population variance (σ^2) is the variance of the sample S^2, which must be calculated using the sample mean $\tilde{X}$. $\tilde{X}$ has been determined after sampling is completed, so as soon as $n-1$ numbers in a sample of size n are determined, the value of the nth number has only one value that makes the sample conform to $\tilde{X}$. In other words, only $n-1$ numbers in the sample are free to vary, and as long as these $n-1$ numbers are determined, the variance S^2 is also determined. Therefore, the mean $\tilde{X}$ is equivalent to a restriction, and due to the addition of this restriction, the sample variance S^2 has a degree of freedom of $n-1$".[7] The reason why the sample variance is to be divided by $n-1$ emanates from this perspective.

4.3.2 The Central Limit Theorem

The theoretical basis of statistical sampling lies in probability theory, and the so-called central limit theorem is a good example of sampling that would otherwise have no theoretical basis. The "common" statement of the central limit theorem (Bennett et al., 2016) is that, in any distribution, a random variable is selected for a number of samples of capacity n. Then, in the case of a large sample (n>=30), the mean distribution is approximately normal; the mean of the mean distribution is approximately equal to the population mean μ; the standard deviation of the mean distribution is $\sigma/n^{1/2}$, where σ is the population standard deviation. Note that it is not the "mean of a sample" but the "mean of the means of the samples" that converges to the population mean.

Here is a more interesting example to illustrate this. It is called the "pegboard experiment"[8], which was proposed by Galton (1822–1911), a cousin of Darwin. "When a small ball touches any nail during its fall, it will roll to the left with 1/2 probability and to the right with 1/2 probability. This continues repeatedly until the ball falls into the grid of the bottom board. The test shows that as long as there are enough balls, the shape of their pile in the bottom plate will approximate a bell-shaped Gaussian curve". This is a good illustration of the central limit theorem: "Each time a ball hits n pegs during its fall, it is equivalent to a 'coin flip' type of random variable. That is, the process of a ball going from the top to the bottom is equivalent to the sum of n coin flips....... It is easy to see that the probability of this average falling in the center is the greatest, but it may also deviate to the left or right by 1 or 2 frames...... The larger the deviation, the smaller the number of balls, and the number of balls at different locations forms a 'distribution', and the Central Limit Theorem proves mathematically that the limit of this distribution is a Gaussian distribution". For a more rigorous proof of the central limit theorem itself, see Appendix III.

4.4 RELIABILITY ESTIMATOR

Assume that the distribution of a particular population is arbitrary. From this population, a sample of size n is drawn, resulting in n observations. These observations are sorted in ascending order:

$$x_1 < x_2 < \dots x_i < \dots < x_n \tag{4.4}$$

where i is the "ordinal number" of observations sorted from smallest to largest. Now assuming that the population distribution probability density is f(x), then the destruction rate $F(x_i)$ of x_i can be determined.

The mathematical expectation of the destruction rate at x_i is shown (Gao, 1986) as the population damage rate valuation:

$$\hat{D} = i/(n+1) \tag{4.5}$$

This is also known as the "average rank". In engineering, the average rank of a sample is often used as the estimate value of population destruction rate.

The estimation of the population reliability at x_i is:

$$\hat{P} = 1 - i/(n+1) \tag{4.6}$$

To prove (4.5), the value of the random variable X_i can be set to x_i. $F(X_i)$ is the destruction rate D of the random variable, and now, we find the mathematical expectation of $F(X_i)$. To do this, let's set that A is an event such as "each sample always has an observation to obtain: $x_i \sim x_i + dx_i$, i.e., $i-1$ observations is less than x_i; the rest of the $n-i$ observations are greater than x_i". There are n! possibilities in total, but the left side of x_i can have $(i-1)!$. Repeated possibilities and $(n-i)!$ repetition possibilities can be obtained:

$$M = n!/[(i-1)! \ (n-i)!] \tag{4.7}$$

This can also be obtained in terms of permutations of M. Consider only x_i and its left element, of which there are i in total, and whose number of combinations is C_n^i. But consider the possibility that there are always i elements as x_i, thus:

$$M = i * C_n^i = i * n!/\left[i!(n-i)! = n!/\left[(i-1)!(n-i)!\right]\right] \tag{4.8}$$

According to the definition of random variable X_i, the probability of occurrence of anything less than x_i is $F(x_i)$, while the probability of occurrence of anything greater than x_i is $(1 - F(x_i))$, which leads to the probability of occurrence of event A is:

$$P(A) = \{n!/[(i-1)! \ (n-i)!]\} \ F^{(i-1)}(x_i)[1 - F(x_i)]^{(n-i)} d[F(x_i)] \tag{4.9}$$

Notice that P(A) is the probability that X_i is between xi and $x_i + dx_i$, and of course, the probability that F(X) obtains between $F(x_i)$ and $F(x_i) + d[F(x_i)]$. So, the mathematical expectation of $F(X_i)$ is:

$$E[F(X_i)] = \int_0^1 F(x_i) \left\{ \{ n!/[(i-1)!(n-i)!] \} F^{(i-1)}(x_i)[1 - F(x_i)]^{(n-i)} d[F(x_i)] \right\}$$

And,

$$E[F(X_i)] = \{ n!/[(i-1)!(n-i)!] \} \int_0^1 \left\{ F(x_i)^i [1 - F(x_i)]^{(n-i)} d[F(x_i)] \right\} \quad (4.10)$$

Noting the definition of the beta function (3.42),

$$B(m, n) = \int_0^1 x^{m-1} (1-x)^{(n-1)} dx$$

therefore:

$$\int_0^1 \left\{ F(x_i)^i [1 - F(x_i)]^{(n-i)} d[F(x_i)] \right\} = B(i+1, n-i+1)$$

However, n, i are positive integers by equation (3.47); so,

$$B(i+1, n-i+1) = [i!(n-i)!]/(n+1)!$$

By equation (4.10), it can be obtained that:

$$E[F(X_i)] = \{ n!/[(i-1)!(n-i)!] \}* [i!(n-i)!/(n+1)!] = i/(n+1) \quad (4.11)$$

which gives equation (4.5), $\hat{D} = i/(n+1)$. And since $\hat{P} + \hat{D} = 1$, equation (4.6) is obtained.

Using equation (4.6) again, the population median estimate can be derived, at which point it can be set: $\hat{P}_{50} = 50\%$

So:

$$1/2 = 1 - i/(n+1) \text{ and } i = (n+1)/2 \quad (4.12)$$

It follows that when n is an odd number, i represents exactly that ordinal number in the center of a set of observations. This makes sense.

It must be emphasized that this reliability estimate is valid for any distribution, and the significance of this conclusion for fatigue reliability estimation is extraordinary. Of course, it is not only for fatigue reliability but for reliability in any sense of the word that it is a benchmark. This is also an important theoretical basis for estimating the three parameters of the Weibull distribution.

4.5 STATISTICAL INFERENCE

4.5.1 Origin of Statistical Inference

Darwin's cousin Galton was a renowned biologist, psychologist, and statistician. His disciple and famous statistician Pearson commented that he "knew more about mathematics and physics than 9 out of 10 biologists, more about biology than 19 out of 20 mathematicians, and more about disease and deformed children than 49 out of 50 biologists".[9] Galton noted (Bennett et al., 2016) that "some people don't like statistics, but I find it full of beauty. No matter when, statistics are not unreasonable. It can explain problems in a detailed and rigorous way if the right methods are used in practical applications. The ability of statistics to deal with complex events is unparalleled". So, is statistics really that "useful"? The answer is yes. Statistical inference is a well-established method that "extrapolates from a sample to population. The population is reflected by the quantitative characteristics of the population distribution, i.e., the parameters (e.g., expectation and variance). Thus, statistical inference includes: estimation of unknown parameters of the population; examination of hypotheses about the parameters; forecasting of the population, etc. The samples used for scientific statistical inference are usually obtained by random sampling methods. The theoretical and methodological basis of statistical inference is the theory of probability and mathematical statistics".[10] The remaining part of this chapter is about learning how to implement statistical inference.

4.5.2 Statistical Definitions of Concepts such as Significance and Confidence

In reality, the following situation is often encountered. For example, Gao (1986) suppose that the logarithm of the fatigue life of a part is normally distributed and the population mean and standard deviation are $\mu_0 = 5.8$ and $\sigma_0 = 0.16$, respectively, based on empirical data. Eight samples are taken from newly manufactured parts, and the mean of the logarithm fatigue life $\tilde{X} = 5.7$ is obtained. Are these parts "qualified"? Or is there a "significant difference" between the two mean values? How to make a decision? This is related to the definition of "qualified". That is, the acceptable "error standard". For this purpose, according to the definition of the standard fraction of the sample:

$$u=(x_p-\mu_0)/s \rightarrow u=(\tilde{X}-\mu_0)/(\sigma_0/n^{1/2}) \tag{4.13}$$

and,

$$u=(5.7-5.8)/(0.16/8^{1/2})=-1.768 \tag{4.14}$$

So, what does this result tell us? Does this batch of parts pass? You can see Figure 4.1:

This figure shows that according to the central limit theorem, the sample mean should be a normal distribution. For $\alpha=5\%$, the standardized acceptance interval is (−1.96, 1.96), so the mean of the sample falls within the acceptance interval, meaning that the sample "qualified". In this sense, on the contrary, if it does not fall in the "acceptance interval", it should be considered as "unqualified". Obviously, the larger α is, the smaller the acceptance interval is; the smaller the "error" is, and the more stringent the requirement of "qualified" is. For this reason, α is defined as "significance level" (hereafter referred to as significance) (Gao & Xiong, 2000).

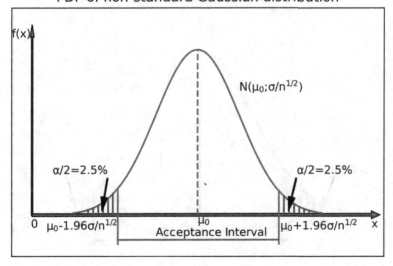

FIGURE 4.1   PDF of non-standard Gaussian distribution.

For convenience, it is generally necessary to standardize it by equation (4.13), so that Figure 4.1 can be changed into Figure 4.2:

Returning to equation (4.14), it can be seen that if the significance is 5%, u is within the acceptance interval (−1.96, 1.96) and therefore can be considered as "qualified", "acceptable", or "no significant difference". However, if the significance α is 10%, the acceptable interval is (−1.64, 1.64), which means that u does not fall within this interval, so it is considered to be "unacceptable" (Figure 4.3).

The larger the α is, the easier it is to determine whether there is a significant difference. So, how do you choose the size of α, i.e., significance, in a practical situation? In fact, there is no uniform standard, it is to be determined according to the specific situation, but if there is no special general, take 5%.

The actual situation is that the population mean value μ_0 is often not known, and one can only make an estimate of it. For this reason, it is first assumed to be an unknown μ, thus equation (4.13) can become that $u = (\tilde{X} - \mu) / (\sigma_0 / n^{1/2})$. Therefore, it can be obtained that:

$$-1.96 < (\tilde{X} - \mu) / (\sigma_0 / n^{1/2}) < 1.96$$

and,

$$\tilde{X} - 1.96 (\sigma_0 / n^{1/2}) < \mu < \tilde{X} + 1.96 (\sigma_0 / n^{1/2})$$

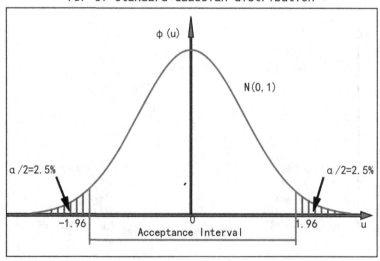

FIGURE 4.2   PDF of standard Gaussian distribution.

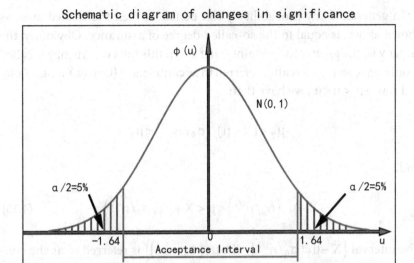

FIGURE 4.3   Schematic diagram of changes in significance.

Noting here that $\tilde{X}$ is related to the sampling of the sample and it is uncertain, therefore, the interval $\left(\tilde{X}-1.96\left(\sigma_0/n^{1/2}\right),\ \tilde{X}+1.96\left(\sigma_0/n^{1/2}\right)\right)$ is also uncertain, i.e., random. However, the probability that this random interval contains μ is 95% (Figure 4.4).

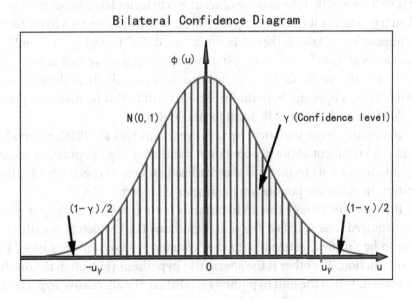

FIGURE 4.4   Schematic diagram of the two-sided confidence.

In general, if any probability γ can be taken, then the shaded area, as shown above, is equal to the so-called degree of assurance. Obviously, the larger γ is, the greater the certainty that μ is in this interval. Then, γ is called "confidence level" (hereafter referred to as confidence) (Gao & Xiong, 2000).

From $-\mu_\gamma < u < \mu_\gamma$, we have that:

$$-\mu_\gamma < (\tilde{X}-\mu)/\left(\sigma_0/n^{1/2}\right) < \mu_\gamma$$

and,

$$\tilde{X}-\mu_\gamma\left(\sigma_0/n^{1/2}\right) < \mu < \tilde{X}+\mu_\gamma\left(\sigma_0/n^{1/2}\right) \tag{4.15}$$

The interval $\left(\tilde{X}-\mu_\gamma\left(\sigma_0/n^{1/2}\right),\ \tilde{X}+\mu_\gamma\left(\sigma_0/n^{1/2}\right)\right)$ is referred to as the confidence interval. It is also easy to see that the larger the confidence level γ, the larger the confidence interval. At the same time, it is easy to see from the above two figures that there is a "complementary relationship" between significance and confidence as follows:

$$(1-\gamma)/2 = \alpha/2 \text{ and } \gamma+\alpha=1 \tag{4.16}$$

4.5.3 Null Hypothesis and Alternative Hypothesis

It is well known that the essence of science is to formulate a hypothesis and then try to falsify it (recognizing it before it is falsified), and if a hypothesis is impossible to falsify, then it is not a "scientific" hypothesis but only a "theological truth". Now, from a statistical perspective, we look at whether a hypothesis can be falsified. If falsified, it means the hypothesis is not valid, and its opposite hypothesis becomes valid. This involves the principle of the excluded middle from formal logic.

According to the statistical perspective (Bennett et al., 2016), a hypothesis is a statement about a population parameter (e.g., population mean μ), and hypothesis testing is a standardized process to check whether the statement about the population parameter is correct.

It is essential to note that in hypothesis testing, at least two hypotheses are required: one is called the null hypothesis (H_0), which is usually the one to be falsified, equivalent to the "counter-proposition" in a proof by contradiction. The other is the alternative hypothesis (H_a), often the one to be proven. When the null hypothesis is falsified, the alternative hypothesis naturally becomes valid.

The outcome of the test of a hypothesis is determined by comparing the actual sample statistic (e.g., the mean) with the result expected by assuming the original hypothesis is correct. To make the decision, a significance level α must be chosen (Bennett et al., 2016):

1. If the probability of obtaining the sample statistic is less than 1% (or 0.01), then the test is statistically significant at the 0.01 level and provides sufficient grounds to reject (falsify) the original hypothesis.

2. If the probability of obtaining the sample statistic is less than 5% (or 0.05), then the test is statistically significant at the 0.05 level and provides a moderately strong reason for rejecting (falsifying) the original hypothesis.

3. If the probability of obtaining the sample statistic is higher than the chosen level of significance (0.01 or 0.05), then we cannot reject the original hypothesis, i.e., it cannot be falsified, and therefore, the alternative hypothesis cannot be accepted.

This testing principle is consistent with Karl Popper's falsifiability concept in philosophy of science, as well as the legal principle of "innocent until proven guilty".

4.5.4 Examples

The statements in the previous section seem to be rather abstract, so let's illustrate them with an example.

Example 4.2

Is a coin "fair" (Bennett et al., 2016)? Suppose someone flips the same coin 100 times and gets 40 heads and 60 tails. Can we conclude that this coin is unfair? (Table 4.2)

Solution

According to the previous theory, we start with a "null" hypothesis, assuming that the coin is "fair", represented by H_0: p=0.5.

TABLE 4.2   Data Using for Example 4.2

p=	0.5	$\sigma=[p(1-p)]^{1/2}=$	0.5
p^=	0.4	$u=(p^\wedge-p)/(\sigma/n^{1/2})=$	-2
n=	100	$u_p=$ norm.s.dist(u,true)=	0.0228

The alternative hypothesis, H_a, is that "the coin is unfair", represented as $p \neq 0.5$. We set the significance level to 0.05. Using Excel, we can obtain Table 4.2.

In the future, it is essential to get accustomed to using "smart tools" like Excel or Python to solve real-world problems. It's apparent because $u_p = 0.0228 < 0.05$; this means that at this significance level, we should reject the null hypothesis and accept the alternative hypothesis because the probability of a "fair" coin producing such results is too low. However, it's important to note that if we choose a significance level of 0.01, then $0.0228 > 0.01$. In this case, we can accept the null hypothesis. This demonstrates that reducing the significance level increases the acceptance interval. It does not imply excessive arbitrariness in statistics but rather highlights that "every truth is context-dependent", and there are no "absolute standards". It's similar to product quality standards, which depend on their application. If used in military or medical contexts, the standards for quality need to be higher; otherwise, slightly lower standards may be acceptable.

By the way, this test is actually u-Test of the normal distribution (Gao, 1986).

Example 4.3: Graduation Salary of University Students (Bennett et al., 2016)

Chinese universities often like to publish the employment rates of their graduates, while American universities prefer to disclose the starting salaries (annually) of their graduates. One year, Columbia University in the United States claimed that the starting salary of its graduates was $39,000. An independent organization called the Advertising Truth Committee suspected exaggeration and decided to conduct a hypothesis test to find evidence to support this suspicion. Following the method from the previous section, they selected H_0: $\mu = \$39,000$ and H_a: $\mu < \$39,000$. Clearly, this is a left-sided test. The committee randomly surveyed $n = 100$ recent Columbia University graduates and obtained $\tilde{x} = \$37,000$ and a sample standard deviation $s = \$6,150$. They used this information to replace σ. You can also solve this problem using Excel (Table 4.3).

TABLE 4.3   For Example 4.3

$\mu=$	$39,000	$\sigma \approx s$	6,150
$\mu^{\wedge}=$	$37,000	$u = (\mu - \mu^{\wedge})/(\sigma/n^{1/2}) =$	3.252
$n=$	100	$u_p = \text{norm.s.dist}(u, \text{true})$	6.E-04

This indicates that at this moment, $u_p = 0.0006$ is not only less than 0.05 but also less than 0.01. This means that the original claim of a population mean $\mu = \$39{,}000$ is highly unlikely given this sample. Therefore, the null hypothesis is rejected, and the alternative hypothesis is supported, indicating that Columbia University did indeed exaggerate.

APPENDIX I: A PROOF OF THE LAW OF LARGE NUMBERS

To prove the law of large numbers, we first need to establish Chebyshev's inequality:

$$P(|X - E(X)| \geq k\sigma) \leq 1/k^2 \tag{A1.1}$$

This is a special case of the following inequality:

$$P(Y \geq K) \leq E(Y)/K \tag{A1.2}$$

where Y can only take non-negative values. We will prove the above inequality (A1.2) because:

$$E(Y) = \int_0^\infty yf(y)dy = \int_0^K yf(y)dy + \int_K^\infty yf(y)dy$$

$$\geq \int_K^\infty yf(y)dy \geq K \int_K^\infty f(y)dy = KP(Y \geq K)$$

This proves (A1.2).

Since Y is non-negative, we can set:

$$Y = (X - E(X))^2, \text{ and } E(Y) = E[(X - E(X))^2] = \sigma^2$$

Setting $K = k^2\sigma^2$, we have:

$$P[(X - E(X))^2 \geq k^2\sigma^2] \leq 1/k^2 \tag{A1.3}$$

This inequality (A1.3) is equivalent to (A1.1). By choosing $k = 3$, we have:

$$P[|X - E(X)| \geq 3\sigma] \leq 1/9$$

This means that the probability within three standard deviations accounts for 8/9 of the total probability. Next, we prove the Bernoulli law of large numbers (Fitz, 1978):

Assuming:

$$P(Y_n = r/n) = C_n^r p^r (1-p)^{n-r} \qquad (A1.4)$$

and,

$$X_n = Y_n - p, \text{ where } 0 < p < 1, r = 0, 1, \ldots, n \qquad (A1.5)$$

Now, we want to prove that:

$$\lim_{n \to \infty} P(|X_n| > \varepsilon) = 0 \qquad (A1.6)$$

We can see that:

$$E(X_n) = 0, \ \sigma^2 = D(X_n) = p(1-p)/n \qquad (A1.7)$$

We can use Chebyshev's inequality as follows:

$$P(|X_n| > k\sigma) \leq 1/k^2 \qquad (A1.8)$$

Since k is arbitrary, we set:

$$k = \varepsilon/\sigma, \text{ i.e., } P(|X_n| > \varepsilon) \leq [p(1-p)]/(n\varepsilon^2) < 1/(n\varepsilon^2) \qquad (A1.9)$$

Therefore, the Bernoulli law of large numbers (A1.6) is proven.

APPENDIX II: UNBIASED AND BIASED ESTIMATES

Here, we further explain why the "unbiased estimate" of the population variance is not the sample variance. Based on the definition of sample variance, we have:

$$S^2 = (1/n) \sum_{i=1}^{n} \left(x_i - \tilde{X}\right)^2 \qquad (A2.1)$$

and,

$$E\left(S^2\right) = E\left[(1/n) \sum_{i=1}^{n} \left(x_i - \tilde{X}\right)^2\right] = E\left[(1/n) \sum_{i=1}^{n} \left(x_i^2\right) - n\left(\tilde{X}\right)^2\right]$$

$$= (1/n)\left\{\sum_{i=1}^{n} E\left(x_i^2\right) - nE\left[\left(\tilde{X}\right)^2\right]\right\} \qquad (A2.2)$$

Noting that: $\tilde{X}=(1/n)\sum_{i=1}^{n} x_i$, and[11] $E\left[(\tilde{X})^2\right]=D(\tilde{X})+E^2(\tilde{X})=\sigma^2/n+\mu^2$; therefore, equation (A2.2) becomes:

$$E(S^2)=(1/n)\left[\left\{\sum_{i=1}^{n}(\sigma^2+\mu^2)-(\sigma^2/n+\mu^2)\right\}\right]=\sigma^2(n-1)/n \qquad (A2.3)$$

The original S^2 is multiplied by a factor $n/(n-1)$, i.e., meaning that equation (A2.1) becomes:

$$s^2=(n/(n-1))S^2=[1/(n-1)]\sum_{i=1}^{n}(x_i-\tilde{X})^2 \qquad (A2.4)$$

Also note that from equation (A2.3), we have:

$$E(s^2)=(n/(n-1))E(S^2)=\sigma^2 \qquad (A2.5)$$

Therefore, $s^2=[1/(n-1)]\sum_{i=1}^{n}(x_i-\tilde{X})^2$ is an "unbiased estimate" of the population variance.

APPENDIX III: A PROOF OF DE MOIVRE-LAPLACE CENTRAL LIMIT THEOREM

The proof of the de Moivre-Laplace limit theorem is as follows (Fits, 1978).
Assuming:

$$P(X_n=r)=C_n^r p^r q^{n-r}, \; 0<p<1, \; q=1-p \qquad (A3.1)$$

and:

$$Y_n=(X_n-np)/(npq)^{1/2} \qquad (A3.2)$$

Then, we have:

$$\lim_{n\to\infty} F_n(y)=[1/(2\pi)^{1/2}]\int_{-\infty}^{y} \exp(-y^2/2)dy \qquad (A3.3)$$

This is known as de Moivre-Laplace limit theorem. It shows that a sequence of standard random variable obeying the binomial distribution converges to a Gaussian distribution, showing that the Gaussian distribution is a limiting distribution of the binomial distribution, revealing the deep intrinsic connection between the two.

In fact, it is known from (Fisz, 1978) that the characteristic function of X_n is that:

$$\varphi_x(t) = (q + pe^{it})^n \qquad (A3.4)$$

and,

$$\varphi_y(t) = \exp[-npit/(npq)^{1/2}]\{q + p\exp[it/(npq)^{1/2}]\}^n$$

$$= \{q\exp[-ipt/(npq)^{1/2}] + p\exp[iqt/(npq)^{1/2}]\}^n \qquad (A3.5)$$

Then, expand the two terms in the right curly brackets according to exp(x), combine them and remove the higher terms, we have that:

$$\varphi_y(t) = (1 - t^2/2n)^n, \text{ i.e.: } \lim_{n\to\infty}\ln(\varphi_y(t)) = -t^2/2. \text{ So, } \lim_{n\to\infty}\varphi_y(t)$$

$$= \exp(-t^2/2) \to \text{q.e.d.}$$

This theorem is very important for practical problems and allows one to estimate the probability of a certain interval. For example, let that:

$$U_n = X_n/n, \text{ because } E(U_n) = p, \text{ Var}(U_n) = pq/n, \text{ therefore,}$$

$$Z_n = (X_n/n - p)/(pq/n)^{1/2} = (U_n - np)/(npq)^{1/2} = Y_n$$

Extending this to the general case, we have that the population sample mean and standard deviation are μ and σ, when $n \to \infty$, $E(\tilde{X}) = \mu$, $\text{Var}(\tilde{X}) = \sigma^2/n$ from (Fitz, 1978).

This is the so-called "generalized" central limit theorem.

Now, we would like to point out the similarities and differences between the law of large numbers and the central limit theorem. From the above proof process, we can see that the law of large numbers is the basis of the central limit theorem. The law of large numbers is mainly concerned with the limit distribution of the mean, while the central limit theorem not only gives the distribution of the mean but also the distribution of the standard deviation, so it is more significant. As can be seen later, the law of large numbers links the basic concepts of probability and statistics, while the central limit theorem gives the theoretical basis for statistical evaluation, which can be said to be indispensable for a wide range of applications in statistics.

Example A.1

Flipping a coin. Suppose a coin is flipped 100 times and ask the probability of the number of times k ($50 < k < 60$) that it comes up heads?

In fact, one can set $\{X_k\}$ ($k = 1, 2, \ldots, 100$), so: $E(X_k) = np = 50$, $Var(X_k) = npq = 25$

Using de Moivre-Laplace limit theorem:

$$P(50 < X_k < 60) = P\left\{\left[(50 - E(X_k))/(npq)^{1/2}\right] < \left[(X_k - E(X_k))/(npq)^{1/2}\right]\right.$$

$$\left. < \left[(60 - E(X_k))/(npq)^{1/2}\right]\right\}$$

$$= P\left\{0 < \left[(X_k - E(X_k))/(npq)^{1/2}\right] < 2\right\}$$

$$\approx \left[1/(2\pi)^{1/2}\right]\int_0^2 \exp(-z^2/2)dz = 0.477$$

NOTES

1 Refer to https://zhidao.baidu.com/question/494702399.html
2 Refer to https://baike.baidu.com/item/%E5%8D%A1%E5%B0%94%C2%B7%E7%9A%AE%E5%B0%94%E9%80%8A/5650305?fr=ge_ala
3 Refer to https://baike.baidu.com/item/%E6%A6%82%E7%8E%87/828845?fromModule=lemma-qiyi_sense-lemma
4 Refer to https://www.jianshu.com/p/c4c90b92d091
5 Refer to https://www.jianshu.com/p/c4c90b92d091
6 The degree of freedom in physics "is the minimum number of coordinates needed to determine the position of a system in space", refer to https://baike.baidu.com/item/%E8%87%AA%E7%94%B1%E5%BA%A6/13237415?fromModule=lemma_search-box
7 Refer to https://www.jianshu.com/p/f012bd5b7e63
8 https://baike.baidu.com/tashuo/browse/content?id=66051d88626e5cd29adc9a3a&lemmaId=829451&fromLemmaModule=list
9 Refer to https://baike.baidu.com/item/%E5%BC%97%E6%9C%97%E8%A5%BF%E6%96%AF%C2%B7%E9%AB%98%E5%B0%94%E9%A1%BF?fromtitle=%E9%AB%98%E5%B0%94%E9%A1%BF&fromid=10239032&fromModule=lemma_search-box

10 Refer to https://baijiahao.baidu.com/s?id=1716477338348458344
11 Refer to https://blog.csdn.net/qq_16587307/article/details/81328773

REFERENCES

Bennett J, Briggs WL & Triola MF (2016), *Statistical Reasoning for Everyday Life*, Posts & Telecom Press, Beijing, p. 149, 127, 222, 223, 225, 227.

Fisz M (1978), *Wahrscheinlichkitsrechnung und Mathematische Stastistik*, Shanghai Science and Technology Press, Shanghai, p. 163, 175, 119.

Gao ZT (1986), *Fatigue Applied Statistics*, National Defense Industry Press, Beijing, pp. 120–124, 226.

Gao ZT & Xiong JJ (2000), *Fatigue Reliability*, Beihang University Press, Beijing, p. 190, 194.

II

Computer Basics of Intelligent Statistical Fatigue

CHAPTER 5

Application of Excel in Probability Statistics

5.1 INTRODUCTION TO EXCEL

Excel is a software in Microsoft Office. According to the relevant information on the Internet, the earliest version of Excel was published more than 30 years ago in 1985 and was used for tabulation, when there were few PCs, so there were few users. It can be said that Excel had been developed along with PC. By the 21st century, PC had been gradually replaced by Notebook, and Excel had also developed significantly. Microsoft has published Excel 2000, 2003, 2007, 2010, 2013, 2016, 2019, and other versions in this century. The use of Excel has greatly increased, and it can complete a number of tasks such as form input, statistics, analysis, etc. and can generate beautiful and intuitive tables and charts. It provides a good tool for dealing with all kinds of forms in daily life.[1] Of course, we are interested in the various functions and graphing capabilities prepared for the statistics, to the general statistics of "small problems" to provide a convenient tool.

This book is not a manual for using Excel, nor a textbook for learning Excel. It only introduces our skills and considerations in the process of using Excel for the reader's reference. The reader who uses Excel to solve the practical problems in the statistics of fatigue should not feel confused.

DOI: 10.1201/9781003488477-7

5.2 POWERFUL FUNCTIONS AND GRAPHING CAPABILITIES IN EXCEL

5.2.1 Excel Functions in Probability Statistics

Excel has hundreds of functions applied in probability and statistics. Even Excel experts cannot master them all. However, for readers of this book, it is important to know that Excel includes almost all elementary functions and commonly used non-elementary functions. Examples include Bessel functions, gamma functions, and beta functions. Generally, any function you want to use can be found in Excel, forming the foundation for intelligent statistics. You no longer need to refer to tables; Excel is sufficient. The challenge lies in knowing how to use them. Smart readers may consult the "Formulas" column. However, upon opening this column, you may find it overwhelming. A good approach is to seek help online, as nearly any question can be answered. Nevertheless, readers still need some basic knowledge, such as functions in Excel being case-insensitive. Additionally, understanding the characteristics of Excel functions and knowing commonly used functions is essential. The following introduces some important and commonly used functions, which are often very similar or identical to functions in programming languages. It's hard to determine who copied from whom.

1. **Truncates Integer Function: INT**

    Syntax: =int(number). Enter "=i" in a cell, and Excel will display all formulas starting with "i" for you to choose from. You can ignore this and continue typing "int(" and then use the cursor to select the cell containing the number you want to round or directly enter a decimal number. For example, to round the number 12.354 in cell A3, use int(A3), and you'll immediately get a return value of 12. Note that this rounding function is "rounding down"; for example, int(–12.354)=–13.

2. **Rounding Function: ROUND**

    Syntax: =ROUND(number, num_digits). This function has two parameters: the number to round and the number of digits after the decimal point. For example, round(12.254,0)=12; round(12.254,1)= 12.3; round(12.254,2)=12.25; round(12.254,–1)=10, where "-1" means "one digit before the decimal point", rounding the units place.

3. **Conditional Function: IF**

    Syntax: =IF(logical_test,value_if_true,value_if_false). The IF function displays the first parameter if the logical expression is true and the second parameter if false. For example, =IF(A5>=650",Excell ent","Average"), where A5 represents someone's total score.

4. **Average Function: AVERAGE**

Syntax: =AVERAGE(number1, number2,...). Returns the arithmetic average of the specified numbers. For example, AVERAGE(1,2,3,4,5,6,7)=4. In practice, you often use the cursor to "highlight" the range of numbers to be averaged. Similar functions include GEOMEAN and HARMEAN.

5. **Sample Standard Deviation Function: STDEV.S**

Syntax: =STDEV.S(number1, number2,...). Returns the standard deviation of these (as a sample) data. It should note that it differs from STDEV.P in the denominator, which is $n-1$ for STDEV.S and n for STDEV.P. For example, STDEV.S(1,2,3,4,5,6,7)=2.16, STDEV.P(1,2,3,4,5,6,7) =2.

6. **Correlation Coefficient Function: CORRE**

Syntax: =CORREL(array1, array2). It gives the correlation coefficient of two arrays. For example, CORREL({1,2,3,4,5},{2,4,7,8,15})= 0.954; while CORREL({1,2,3,4,5},{9,7,4,3,2})=−0.976. In practice, array1 and array2 are often entered by "highlighting" a row or column in the worksheet.

7. **Covariance Function: COVAR**

Syntax: =COVAR(array1, array2). That is, the mean of the product of the deviations of each pair of variables. Pay attention to its relationship with the correlation coefficient. Using the previous example, COVAR({1,2,3,4,5},{2,4,7,8,15})=6 but STDEV.P(1,2,3,4,5)=1.414; STDEV.P (2,4,7,8,15)=4.445 so, 6/(1.414*4.445)=0.954.

8. **Slope of Linear Regression Function: SLOPE**

Syntax: =SLOPE(known_y, known_x). It gives the slope of the regression line for the given data points. Note that the first parameter is the y-coordinates of the data points, and the second parameter is the x-coordinates. The order matters. For example, SLOPE({2,4,7,8,15}, {1,2,3,4,5})=3, while SLOPE({1,2,3,4,5},{2,4,7,8,15})=0.303.

9. **Linear Regression Intercept Function: INTERCEPT**

Syntax: =intercept(known_y, known_x). It gives the intercept of the regression line for the given data point. Similarly, the order matters. For example, intercept ({2,4,7,8,15},{1,2,3,4,5})=−1.8 and intercept ({1,2,3,4,5},{2,4,7,8,15})=0.814

10. **Gamma Function: GAMMA**

Its syntactic structure is gamma(x), where x is a real number greater than zero. If x is an integer n greater than 1, then gamma(n)=(n−1)! and gamma(0.5)=$\pi^{1/2}$=1.77245.

11. **Standard Normal Distribution Function: NORM.S.DIST**

Syntax:=NORM.S.DIST(z, TRUE), where z is the "standard fraction" of the standard normal distribution. If the second parameter is "TRUE", then the value of the cumulative distribution function CDF; if "false", then the value of the probability density PDF of the point. For example, NORM.S.DIST(1,TRUE)=0.841, while NORM.S.DIST(1,false)=0.242. Thus, even without the normal distribution function distribution table, there is no problem at all. Here, it should be noted that the old version of the standard normal distribution function is NORMSDIST although also available but cannot represent PDF values. That is, the mean of the product of the deviations of each pair of variables. Here, x is not the standard fraction, the second and third parameters are the mathematical expectation of the normal distribution and the standard deviation, the fourth parameter if TRUE, it is the CDF, and if FALSE, it is PDF.

12. **Inverse Function of Standard Normal Distribution: NORM.S.INV**

Syntax: =NORM.S.INV(cumulative probability). This is essentially the inverse function of the standard CDF, giving the standard score for a given cumulative probability. This is crucial in probability and statistics calculations where table lookup is not only cumbersome (and error-prone) but also inaccurate (often requiring estimation). For example, NORM.S.INV(0.8413)=0.999815. The reason it is not exactly equal to 1 is that cumulative probability values are rounded to the ten-thousandth place, so the inverse operation is accurate only to the ten-thousandth place.

13. Many commonly used distribution and test functions are available. Here are some examples:

CHISQ.DIST: the left-tailed probability of the χ^2 distribution.

CHISQ.INV: the inverse function of the left-tailed probability of the χ^2 distribution.

CHISQ.DIST.RT: the right-tailed probability of the χ^2 distribution.

CHIEQ.INV.RT: the inverse function of the right tail probability of the χ^2 distribution.

CHISQ.TEST: χ^2 test results.

F.DIST: the left-tailed probability of the F-distribution.

F.INV: the inverse function of the left-tailed probability of the F-distribution.

F.DIST.RT: the right tail probability of F-distribution.

F.INV.RT: the inverse function of the right tail probability of the F-distribution.

F.TEST: F-test results.

T.DIST: left-tailed t-distribution.

T.DIST.RT: right-tailed t-distribution.

T.DIST.2T: Two-tailed t-distribution.

T.TEST: t-test results

Z.TEST: z-test results.

The listed functions are what the author considers to be useful for probability and statistics, at least meeting the basic requirements for the "intelligent fatigue statistics". For readers, it is essential not only to know how many functions are available in Excel but more importantly, to be able to proficiently apply these functions to solve problems in practical work. Therefore, continuous and conscious use of these functions is necessary; otherwise, just like many learned concepts in school, they will quickly be forgotten.

5.2.2 Some Notes on Excel in Graphing

As Excel continues to undergo continuous updates, its graphing capabilities become increasingly powerful. Here, we won't introduce all the graphing features,[2] but rather, we'll provide examples to illustrate how to create charts in Excel. Suppose we have the fatigue life data (Gao, 1986) for certain components, and we now need to verify whether these data follow a normal distribution:

Let's start with an explanation of this table. The third column, "fatigue life", represents the raw data, and the original order was "random".

After applying Excel's "auto-sort" function, we obtain the sorted data. The second column, "Reliability Estimate" (multiplied by 100), was introduced in Section 4.4. It is calculated directly from the first column using equation (4.6). The fourth column, "logarithm of fatigue life", represents the logarithm of the values in the third column. The Gaussian distribution for fatigue life is actually in terms of the "logarithm of fatigue life". The fifth column corresponds to the Gaussian distribution based on the data in the second column. The "standard score" for this distribution can be obtained using NORM.S.INV(1-reliability*0.01). The sixth column's "theoretical standard score" is calculated based on the mean and sample standard deviation of the Gaussian distribution estimated from the data in the fourth column, following equation (3.9). Therefore, if these data truly follow a Gaussian distribution, the result should be a straight line. Thus, testing whether these data conform to a Gaussian distribution essentially involves examining whether the "standard scores" align with the "theoretical standard scores". Two curves need to be plotted, both using the fourth column as the x-coordinate (Table 5.1).

To do this, you can find the "Charts" option under "Insert" in Excel, open the "Scatter Plot" chart, and customize it as needed to obtain Figure 5.1. The details of creating the chart are not discussed here, and readers interested in this topic can find relevant information online.

Therefore, it can be observed that the logarithm of fatigue life for these components can essentially be regarded as following a

TABLE 5.1   Fatigue Life Data Sheet for a Part

No.	Reliability Est.	Fatigue Life	$x_i = lgN_i$	Stand. Frac.	Theory' Stand. Frac.
1	9.091E+01	**124**	2.09E+00	−1.335E+00	−1.476E+00
2	8.182E+01	**134**	2.13E+00	−9.085E-01	−8.003E-01
3	7.273E+01	**135**	2.13E+00	−6.046E-01	−7.355E-01
4	6.364E+01	**138**	2.14E+00	−3.488E-01	−5.441E-01
5	5.455E+01	**140**	2.15E+00	−1.142E-01	−4.187E-01
6	4.545E+01	**147**	2.17E+00	1.142E-01	6.366E-03
7	3.636E+01	**154**	2.19E+00	3.488E-01	4.116E-01
8	2.727E+01	**160**	2.20E+00	6.046E-01	7.446E-01
9	1.818E+01	**166**	2.22E+00	9.085E-01	1.065E+00
10	9.091E+00	**181**	2.26E+00	1.335E+00	1.819E+00
		Sample' μ=	2.17E+00		
		Sample' σ=	4.98E-02		

Data in bold is the original data, and the rest are calculated based on it.

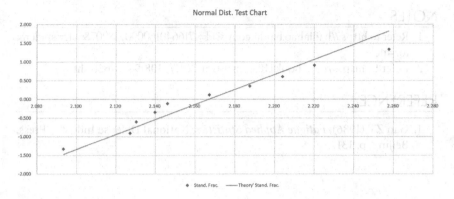

FIGURE 5.1   Normal distribution test chart.

Gaussian distribution. However, this judgment, as indicated in Section 3.1.4, is not strict. This example is merely intended to illustrate a usage of Excel.

Certainly, there are many types of charts available in Excel, and it's not possible to introduce them all here. However, readers should have the concept that using Excel for calculations and plotting is relatively straightforward, especially when dealing with moderate amounts of data, and it proves quite convenient for on-site applications. On the other hand, Python, as introduced in Chapter 6, can overcome limitations where Excel falls short, although it may be comparatively more "complex". Nevertheless, it is easier to learn compared to other programming languages. Whether using Excel or Python, repeated practical application is necessary to enhance proficiency and prevent forgetting.

5.3 EXCEL'S STRENGTHS AND WEAKNESSES

Excel, as a versatile software capable of tasks from spreadsheet creation to statistical analysis and charting, may seem omnipotent. However, in reality, it has its limitations. To illustrate, consider solving transcendental equations using Newton's method. While it is not "absolutely" impossible, it is undoubtedly cumbersome and, most importantly, lacks "intelligence"—the ability to automatically generate the required tables. Therefore, alternative tools must be employed, such as Python, which will be shown to perform tasks that Excel cannot. Despite this, it is crucial to emphasize that specific problems require specific analyses; there is no one-size-fits-all software—only the most suitable software for a given task.

NOTES

1 Refer to https://baijiahao.baidu.com/s?id=17166401600907889078&wfr=spider&for=pc
2 Refer to https://max.book118.com/html/2017/0308/94681628.shtm

REFERENCE

1. Gao ZT (1986), *Fatigue Applied Statistics*, National Defense Industry Press, Beijing, p. 131.

CHAPTER 6

Introduction to Python

6.1 GETTING STARTED WITH PYTHON

6.1.1 The Origin of Python and its Advantages and Disadvantages

Python was only created in the early 1990s as an object-oriented high-level programming language "Python allows developers to express ideas in less code than C++ or Java. Whether small or large, the language attempts to make the structure of a program clear and straightforward....... The Python virtual machine itself can run in almost any operating system. Python's formal interpreter, CPython, is written in C and is community-driven free software".[1] This makes it very user-friendly, and Python's design philosophy is "elegant", "clear", and "simple". The Python developers' philosophy is "one way, preferably only one way, to do something", which is why it is so different from other languages that have a distinctly personal style. When faced with multiple choices when designing a Python language, Python developers generally reject fancy syntax in favor of syntax that has no or little ambiguity. These guidelines are called Python maxims. Simplicity and clarity in writing programs is very much in demand. This also makes it easy for users to grasp and use. In many cases, Python is almost self-explanatory as long as you have good English and a basic knowledge of other computer languages. A particular advantage is that it has many libraries. "The Python standard library is really large, and Python has definable third-party libraries that can be used. It can help you with a variety of tasks, including regular expressions, document generation, unit testing, threading, databases, web browsers, CGI, FTP, email,

DOI: 10.1201/9781003488477-8

XML, XML-RPC, HTML, WAV files, password systems, GUIs (graphical user interfaces), Tk, and other system-related operations. Remember, all of these features are available as long as Python is installed. This is called Python's "full-featured" philosophy. In addition to the standard libraries, there are many other high-quality libraries such as wxPython, Twisted, and the Python Image Library and so on".[2] Of course, Python is not without its drawbacks, and sometimes, the so-called advantages can be disadvantages from a certain point of view. For example, the open source of the code can be confusing for secrecy. There are many libraries that can be called, and this can affect the speed of operation. But the defects cannot obscure the virtues, in fact, more and more people are using Python to write programs. And this book will focus on Python to solve the problems encountered in fatigue statistics.

It's worth noting that Python was the "Programming Language of the Year" in 2018 according to Tiobe (The Importance of Being Earnest). Python received this title because it ranked the highest compared to other languages in 2018. Following Python, Visual Basic.NET and Java ranked second and third, respectively. Python has now become a significant part of large programming languages. For nearly 20 years, C, C++, and Java have consistently been the top three languages, far ahead of other programming languages. Python has now joined these top three languages. It is currently the most commonly taught first programming language in universities, ranking first in the field of statistics, artificial intelligence programming, scriptwriting, system testing, web programming, scientific computing, and more. In summary, Python is ubiquitous.[3] According to online sources, "The 2021 Programming Language Ranking is out, and Python is back on top".[4] Even though Python ranked third in 2022, it is still considered the "best programming language for machine learning".[5]

The authors believe that Python is not only suitable for machine learning and artificial intelligence (AI) but also highly versatile for scientific computing. When it comes to probability and statistics calculations and data visualization, Python is on par with other specialized software such as MATLAB, SPSS, SAS, and R. It can handle all calculations and data visualization tasks in fatigue reliability without any issues. Considering that readers engaged in fatigue reliability may not be familiar with Python, this book just adds this chapter specifically to introduce this essential intelligent tool. However, this book is not a Python textbook and cannot provide a comprehensive introduction to Python. It will focus on providing knowledge related to applications in this field. The authors believe that by learning and using Python, readers will be able to apply it to various areas of interest and gain valuable skills. If readers can achieve this, the authors will be delighted. Additionally, in this "intelligent" era, mastering computer languages is as important as mastering mathematics and languages and is considered a fundamental skill.

6.1.2 How to Build a Platform that Uses Python

With the rapid development of computer hardware and software, computer languages are constantly evolving, and there are so-called runtime environment requirements for using different languages. It is not just a matter of downloading a certain version of a language. This also applies to Python. Some early published books on Python don't properly introduce Anaconda Python Distribution as a tool for using Python. This tool is too handy for Python. It is available at https://www.anaconda.com/ along with the steps for using it.

A special reminder is that Anaconda comes with a very good IDE (integrated development environment), namely Spyder. There are syntax errors that are immediately prompted, copy and paste are very convenient, and debugging procedures are relatively easy. It is considered to be the best IDE for Python programs. Of course, there are also good IDEs like PyCharm, which is suitable for web development to optimize the design, but not much help for the work used in statistics and other aspects. Many people also like to use Jupyter (notebook) and other IDEs, all of which have various features and advantages and disadvantages, not one by one here.

After downloading Anaconda, look for the Anaconda Navigator, Anaconda Prompt, and Spyder startup commands in the Start menu and "pull" them to your desktop for future use.

Anaconda also has the advantage that it comes with a "package manager" called Conda, which is almost identical to the Python-specific manager pip. The main difference is that "pip installs Python packages, while Conda installs packages that may contain software written in any language. Before using pip, the Python interpreter must be installed through the system package manager or by downloading and running the installer. Conda installs Python packages and the Python interpreter directly".[6] The Conda "command" allows you to update, install, etc. the libraries used in Python. However, it is often more convenient to operate on Anaconda itself. Here's an example:

Double-click on the icon of Anaconda ⟳ on the desktop and it will appear (Figures 6.1).

Then, place the cursor on the "Environment" and click it, and the following screen will appear,

At this time, just click the "<" in the middle, and the following screen will appear.

This indicates that 519 packages have been installed in the existing Python, and then pressing the down "∨" next to "Install" will bring up many options, one of which is Updatable. Selecting this option will bring up,

Indicates that there are 71 packages that can be updated. You can choose all the update or the packages you are interested in. It should be noted that due to the network, sometimes it takes a long time to update a certain package, which requires enough patience.

As the application of Python expands, some new packages are often needed. For this reason, you can first look for it in "Not Installed". If you find it, install it immediately. If you can't find it, you can find it in

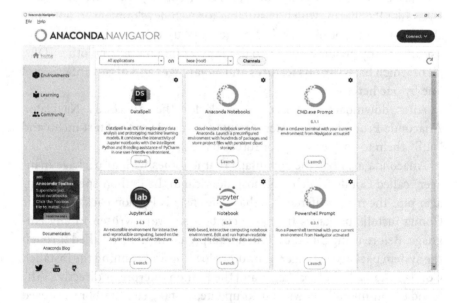

FIGURE 6.1   Anaconda's running interface.

Anaconda Cloud. Here, it should be noted that you must find the version of the hardware suitable for your computer, such as the computer with what operating system and the number of bits of memory. Is it 32 bits or 64 bits?

FIGURE 6.2   Anaconda's environments interface.

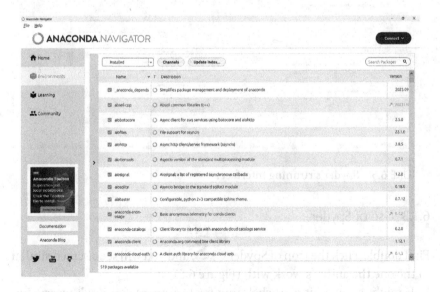

FIGURE 6.3   Anaconda's base(root) interface.

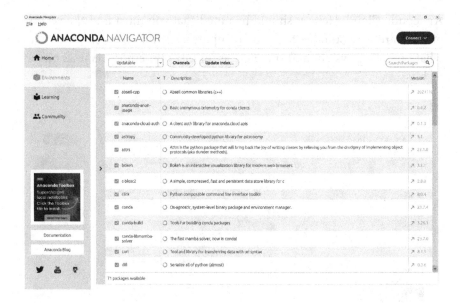

FIGURE 6.4   Anaconda's update interface.

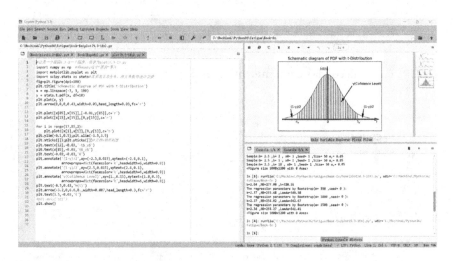

FIGURE 6.5   Spyder's running interface.

6.1.3 Use of Spyder

First double-click the icon of Spyder , and you will see a screen similar to the one the authors work with (Figure 6.5).

It might seem a bit overwhelming for beginners, but don't worry. Just like with smartphones, while they have powerful features, regular users only need to learn how to make calls, take photos, and use WeChat. Modifications to photos and the use of various apps can be learned as needed. The same goes for Spyder; start by learning the most basic operations:

1. **Inputting Code:** There are two methods. First, you can type the code directly in the "Code Editing Area" on the left. Second, you can copy existing code and paste it into the "Code Editing Area" by right-clicking and selecting the "Paste" option. For example, the code in Figure 6.5 was written to draw Figure 9.9 in Chapter 9.

2. **Running the Code:** If Spyder doesn't indicate any errors, it means there are no syntax errors in your code, and you can run it. There are two methods to run the code. First, you can click on the "Run" option in the toolbar and select the "Run" option in the dropdown. Second, you can click on the filled triangle in the second row of the toolbar. The results of running the code will be displayed in the "plot" area in the upper right. If you want to copy this plot to Word, simply move your cursor over the plot (either the larger or smaller one in the upper right) and right-click to find the "Copy" option and then click on it.

3. **Naming and Saving Files:** If the code you entered hasn't been saved as a file beforehand, Spyder will prompt you to name and save the code when you run it. Just like in Microsoft Word, Spyder also has a "Save As" feature. You can access it by clicking on "File" in the toolbar. Other functions in the toolbar will not be explained in detail.

6.2 PYTHON'S BASIC GRAMMAR

Here, we emphasize the characteristics of Python grammar, and we will talk less or not talk about the same parts as other programming languages (Liu et al., 2018). Similar to general linguistics, Python Grammar can be divided into two parts, namely morphology and syntax. It should be noted that the two cannot be completely separated. A "word" must have its meaning revealed by the context of the sentence; likewise a sentence cannot exist without words. But for the convenience of study and narration, they have to be narrated separately. By contrast, "word" must be more basic than "sentence". When learning this grammar, we strongly advise readers to run it on Spyder while reading it, which can get twice the result with half the effort. The morphology and syntax of Python below are written with reference to Lutz (2015).

6.2.1 Python's Morphology

6.2.1.1 Number, Type, and Operation

6.2.1.1.1 Types of Numbers   Numbers are naturally one of the most basic "characters" of all computers. However, due to the nature of computers— limited capacity and inability to represent "continuous" real numbers in a mathematical sense—there is always a precision limitation. However, what sets Python apart from other languages is its dynamic allocation of space, so it seems that it has no limit on the number of digits it can express for integers. In general, overflow does not need to be considered.

In Python, unlike languages such as FORTRAN, C, and JAVA, there is no distinction between single and double precision in floating-point representation; there is only one type of floating-point number. For Python 3.X, it provides 17 significant digits with a range from 10^{-308} to 10^{308}. In Python, integers are automatically converted to floating-point numbers, but the reverse does not happen.

6.2.1.1.2 Number Operations   Numeric operations in Python are represented by "+", "−", "*", and "/". Similar to C and Java, Python also has a modulus operation "%", but unlike C and Java, Python uses "//" to represent integer division, which strictly rounds down the result. Running the following code in Spyder:

```
print('5/3=',5/3)# If it can't be divided, it will automatically be represented by
floating-point numbers.
print('5/3.1=',5/3.1)
print('5//3=',5//3)# just give the (downward) integer quotient
print('-5//3=',-5//3)# just give the (downward) integer quotient
print('5%3=',5%3)# means remainder operation, and it becomes modular operation when the
numerator and denominator are all integers.
print('5%3.1=',5%3.1)# means remainder operation.
```

The result will be:

```
5/3= 1.6666666666666667
5/3.1= 1.6129032258064515
5//3= 1
-5//3= -2
5%3= 2
5%3.1= 1.9
```

6.2.1.2 Strings and Numbers

Having covered numbers, let's discuss strings. It's challenging to provide a strict definition of a string, but you can think of it as a collection of symbols composed of any characters that the computer can support, such as letters, digits, and even Chinese characters. Strings are a data type in computing. It's important to note that strings are also a form of "data" because they can be "digitized", just like sound and images. This is what enables computers to move from purely numerical calculations to more general-ized computations, forming the basis of intelligence.

For individuals who are not professionally involved in computer work, understanding the basic functionality of strings is sufficient. Therefore, here we will only touch on the most fundamental concepts, and more in-depth applications can be found online.

6.2.1.2.1 Generation Strings   There are two methods to generate strings. First, you can enclose the desired character set in single or double quotes. Second, you can use the str() function to convert numeric data into strings. Running the following code in Spyder:

```
a='123'
print('a=',a)
b=str(123.56)
print('b=',b)
d='123+456='
print(d)
e=str('123+456=')
f=123+456
print(e,f)
```

The result will be:

```
a= 123
b= 123.56
123+456=
123+456= 579
```

As you can see, for numeric values, using single quotes or the str() function yields the same result. However, for non-numeric values like '123+567=', using str() to convert it into a string doesn't make much sense, as you cannot remove the single quotes in this case. If you remove the quotes, Spyder will report an error. In this case, it's better to use variable.

6.2.1.2.2 The Relationship between Numbers and Strings   As mentioned ear-lier, numbers can be converted into strings, but not all strings can be con-verted into numbers. So, what's the use of numeric strings? Consider the examples of variables d and f in the previous code. The former is a string representing an arithmetic expression involving two numbers, while the latter is a numeric value representing the result of adding those two num-bers. Only by combining them can you get the final line of the output: "123+456=579."

6.2.1.2.3 Some Basic Operations between Strings   Strings support various operations, including regular addition with "+", special addition with "join", reverse operations with "split", and replacement operations with "replace". Running the following code in Spyder:

```
E='Hello'
F=' World'
print('E+F=',E+F)# Ordinary addition of string
A=['1','2','5','4','3']
B='+'
print("B.join(A)=",B.join(A))# Special addition "join" of string
D=B.join(A)
print ("d.replace ('+','-') =", d.replace ('+','-')) # The'+'of the string is replaced by'-'
        C='1+2+3+4+5'
print ("c.split ('+') =", c.split ('+')) # Splitting strings according to specified rules
```

The result will be:

```
E+F= Hello World
B.join(A)= 1+2+5+4+3
D.replace('+','-')= 1-2-5-4-3
C.split('+')= ['1', '2', '3', '4', '5']
```

6.2.1.3 List

Lists in Python can be seen as a form of linear (data) structure. If all elements inside a list are of the same type, it can be seen as a "(one-dimensional) vector", making it a "generalized" vector. This is the biggest difference between lists and regular vectors, as lists can contain elements of different types. Additionally, lists are characterized by their order and mutability. It will be illustrated below.

6.2.1.3.1 Creating List   There are many ways to create lists in Python. You can use square brackets "[]" to represent an empty list, define specific elements like [1, 2, 3, 4, 5], use the list() function, and even create nested lists, where a list contains sublists, like [1, [2, 3], [4, 5]]. You can also generate a new list using various built-in functions and methods. For example, A.append(X) adds element X to the end of the existing list A, and if A is initially empty, it effectively creates a new list. Similarly, A.extend(B) adds all elements from list B to the end of list A. Running the following code in Spyder:

```
A=[]
A.append(6)
print('A=',A)
B=[1,[2,3],[4,5]]
print('B=',B)
B.extend(A)
print('new B=',B)
```

The result will be:

```
A= [6]
B= [1, [2, 3], [4, 5]]
new B= [1, [2, 3], [4, 5], 6]
```

6.2.1.3.2 List Slicing   List slicing in Python is done using the syntax [left: right] or [left: right: stride]. It's important to note that the positions (indices) of elements in a list start from 0, and the last position is sometimes represented as −1. Assuming the variable A has value [1, 3, 5, 7, 8, 13, 20], the following statements are true:

```
A [2: 5] = = [5, 7, 8]: Slicing from index 2 (actually the 3rd element) up to index 5 (not
including index 5; in Python, it's "left-closed, right-open").
A [1:] = = [3, 5, 7, 8, 13, 20]: Slicing from index 1 to the end.
A [:-3] = = [1, 3, 5, 7]: Slicing from the beginning to the third-to-last element.
A [:] = = [1, 3, 5, 7, 8, 13, 20]: Returns all elements. Modifying the new list won't affect A.
A [1: 5: 2] = = [3,7]: Slicing from index 1 up to index 5 with a step size of 2.
```

6.2.1.3.3 List Addition (Plus) and Multiplication   Only lists of the same type can be added. Running the following code in Spyder:

```
a=[1,2,3]
b=[4,5,6]
c="Hello "
d="World"
print ("a+b=:",a+b)
print ("c+d=:",c+d)
```

The result will be:

```
a+b=: [1, 2, 3, 4, 5, 6]
c+d=: Hello World
```

The so-called multiplication refers to the repetition of elements in a list. For example:

```
a=[1,2]+5*[3]
b=2*[4,5,6]
print ("a=:",a)
print ("b=:",b)
```

The result will be:

```
a=: [1, 2, 3, 3, 3, 3, 3]
b=: [4, 5, 6, 4, 5, 6]
```

6.2.1.3.4 Some Operators or Functions Related to List   Frequently used functions include *len, max, min*. Their functionality is self-explanatory. Running the following code in Spyder:

```
A=[20,10,5,100,-3]
AL=len(A)# indicates the "length" of A, which is also the number of elements in A
Amax=max(A)# that the largest element in A
Amin=min(A) # that the smallest element in A
print ("AL=",AL,", Amax=",Amax,", Amin-=",Amin)
```

The result will be:

```
AL= 5, Amax= 100, Amin-= -3
```

In addition to the *append* and *extend* functions mentioned earlier, there are some commonly used functions for lists, such as *del* for partial deletion, *pop* for removing the last element, *insert* for inserting an element at a specific position, *sort* for sorting, and *reverse* for reversing the order. Here's an explanation using Spyder:

```
users=['peter','jin','york','david','mark']
del users[2]#del the element with subscript 2 (i.e. the third)
print ("users=",users)
B=['b','bb','bbb']
print("B.pop()=",B.pop(),", B=",B)
A=['a','c','f','a','h','w','a']
A.insert(3,'abc')#insert abc at position 3+1=4 in A
C=[3,2,5,1,4]
C.sort()
print ("C.sort=",C)
C.reverse()
print ("C.reverse=",C)
```

The result will be:

```
users= ['peter', 'jin', 'david', 'mark']
B.pop()= bbb, B= ['b', 'bb']
A= ['a', 'c', 'f', 'abc', 'a', 'h', 'w', 'a']
C.sort= [1, 2, 3, 4, 5]
C.reverse= [5, 4, 3, 2, 1]
```

6.2.1.4 Tuple

A tuple is an unchangeable list, meaning once defined, it cannot be modified. Unlike lists, which use square brackets "[]", tuples are represented using parentheses "()". You can create tuples using the tuple() function. For example, running the following code in Spyder:

```
A=['a','b','c']
B=[1,2,3]
C='abc'
print("tuple(A)=",tuple(A) )
print("tuple(B)=",tuple(B) )
print("tuple(C)=",tuple(C) )
print("tuple('abc')=",tuple('abc') )
```

The result will be:

```
tuple(A)= ('a', 'b', 'c')
tuple(B)= (1, 2, 3)
tuple(C)= ('a', 'b', 'c')
tuple('abc') = ('a', 'b', 'c')
```

Interestingly, the results for A and C are the same. From the perspective of tuples, both A, which originally has three elements and is a list, and C, which has only one element and is a string, are considered "the same".

In most cases, people find lists more convenient, so what is the use of tuples? The main advantage of tuples lies in their immutability.

6.2.1.5 Dictionary

A dictionary is a special data structure and the only built-in mapping type in Python. It is highly flexible. Unlike lists, dictionaries do not have a specific order but store data with specific keys. They are represented using curly braces "{}". Here are some code examples in Spyder:

```
D={'a':1,'b':2,'c':3,'d':4,'e':5}
print(list(D.keys()))#Print a list of the keywords in the dictionary D
print(list(D.values()))#Print a list of the values in the dictionary D
print(list(D.items()))#Print a list of key pairs in the dictionary D
print(D)#print out the dictionary D
```

The result will be:

```
['a', 'b', 'c', 'd', 'e']
[1, 2, 3, 4, 5]
[('a', 1), ('b', 2), ('c', 3), ('d', 4), ('e', 5)]
{'a': 1, 'b': 2, 'c': 3, 'd': 4, 'e': 5}
```

Although dictionaries are essential, their primary focus in our context is not on probabilistic statistical calculations and plotting, so we won't delve deeper into them.

6.2.1.6 Variables and Types

If numbers and Latin letters are the "letters" in Python, then variables are the "words". Like variables in mathematics, they can represent any type of data used in a computer.

6.2.1.6.1 Naming Rule   Due to the complexity and diversity of computer applications, it is necessary to follow certain naming rules:

1. Do not use the same keywords as in Python.

2. Use precise English naming, so that the code written like numbers can be understood by everyone.

3. Three commonly used international naming conventions: Camel Case, Pascal Case, and Hungarian Notation.[7]

TABLE 6.1 Variables and Types

prefix	a	b	by	fn	i	s
type	Array	Boolean	Byte	function	Integer	String

These naming rules are essential for professionals in the computer software industry but can also serve as an important reference for non-professionals (Table 6.1).

6.2.1.6.2 Types of Variables Data types can be divided into two types, numeric and non-numeric. The numeric type includes integer (int), floating point (float), complex, and Boolean (bool). The three types have been introduced in the previous sections, now focus on the Boolean type. The so-called Boolean is actually the logic of true and false and in mathematical logic has been "digitalized". In this sense, Boolean types can be regarded as numeric types. Non-numeric types include strings, lists, tuples, and dictionaries. It is worth noting that the so-called non-numeric data are much more than the four types pointed out above, such as matrices, tensors (multidimensional arrays), graphs, and so on.

6.2.1.6.3 Definition and Conversion of Variable An important advantage of Python over C and Java is that there is no need to "define" the type of variable in advance, but this may be a "disadvantage" from another point of view, that is, it is not "rigorous" enough. In Python, the interpreter does this automatically. If it is necessary to change the type of variable, another function needs to be used. When you use a variable, you don't need to declare it in advance, you just need to assign a value to it. Therefore, the type of the variable is determined during the assignment process. The operator "str" is used to convert numbers to strings, but not the inverse operation, which converts strings made up of numbers to numbers. "int" is used to turn an integer string into an integer and a floating-point number into a downward-facing integer, while "float" is used to turn a floating-point string into a floating-point number and an integer number into a floating-point number. Note, however, that not all different data readings can be converted to each other. For example, we run the following code on Spyder:

```
print("int('12')=",int('12'))# convert string '12' to integer
print("int(12.56)=",int(12.56))# convert the floating-point number 12.56 to an integer
print("float('12.34')=",float('12.34'))# convert the string '12.34' to a floating-point
number
print("float(12)=",float(12))# convert the integer 12 to a floating-point number
```

The result will be:

```
int('12')= 12
int(12.56)= 12
float('12.34') = 12.34
float(12) = 12.0
```

6.2.1.7 Python Built-in Functions

Just like in mathematics, there are variables and then there are functions. There are many built-in functions in Python, see https://docs.python. org/zh-cn/3/library/functions.html. Here, we just select some that we think we'll use often. In fact, some of them have already been mentioned above, such as *"str()"*, *"int()"*, *"float()"*, *"print()"*, *"list()"*, *"tuples()"* and *"dict()"*

6.2.1.7.1 Operators   Operators in Python can also be seen as a kind of built-in function. For example, the arithmetic operators we discussed earlier, especially the less common ones like floor division // and modulo %, can be considered as function operations (Tables 6.2–6.4). Here, we will specifically introduce logical operators, comparison operators, and compound assignment operators:

TABLE 6.2   Logic Operators

Logic Oper.	Explanations
and	Equivalent to the "product" of two sets, where both expressions are true and the result is true
or	Equivalent to the "sum" of two sets, where one of the two expressions is true and the result is true
not	Equivalent to the "remainder" of a set, which means that if the expression is true, the result is false, and vice versa

TABLE 6.3   Comparison Operators

Compar. Oper.	Explanation
==	Judge whether the two sides are equal, if the equality is true.
!=	Judge whether the two sides are unequal, if the inequality is true.
>	True if the left is greater than to the right.
>=	True if the left is greater than or equal to the right.
<	True if the right is greater than to the left.
<=	True if the right is greater than or equal to the left.

TABLE 6.4   Compound Assignment Operators

Assign. Oper.	Explanation	For ex. start with i = 2
+=	Assignment is the result of addition.	i+=1→i=3
-=	Assignment is the result of subtraction.	i-=1→i=2
*=	Assignment is the result of multiplication.	i*=19→i=38
/=	Assignment is the result of dividing by the number on the right.	i/=2→i=19.0
//=	Assign integer part of the quotient divided by number on right.	i//=2→i=9.0
%=	Assign the remainder of numb. divided by numb. on right.	i%=6→i=3.0
**=	Assign a value to the power of the number to its right.	i**=4→i=81.0

Let's illustrate this with some examples in Spyder:

```
a=1;b=2;c=3
print((a>b)and(c>b))# two expressions have a false for and is false
print((a<b)and(c>b))# both expressions are true and is true
print((a>b)or(c>b))#Two expressions have a true then for or is true
print(not(a>b))#because at this point a>b is false, so not(a>b) is true
print(not(a!=b)and(c>b))#Two expressions have a false for and is false
```

The result will be:

```
False
True
True
True
False
```

The same can be illustrated by running the following code on Spyder:

```
i=2
print('Initial i=',i)
i+=1# is equivalent to i=i+1, sometimes it can be simplified to i+, if the right side just adds 1
print('after i+=1 operation i=',i)
i-=1# is equivalent to i=i-1, sometimes it can be simplified to i-, if the right side just adds 1
print('after i-=1 i=',i)
i*=19#equals i=i*19
print('After i*=19 i=',i)
i/=2# is equivalent to i=i/2
print('i= after i/=2',i)
i//=2# is equivalent to i=i//2
print('i= after i//=2',i)
i%=6#equals i=i%6
print('i= after i%=2',i)
i**=4#is equivalent to i=i**2, which is pow(i,4)
print('After i**=4 i=',i)
```

The result will be:

```
The initial i= 2
i= 3 after i+=1
i= 2 after i-=1
i= 38 after i*=19
i= 19.0 after i/=2
i= 9.0 after i//=2
i= 3.0 after i%=2
i= 81.0 after i**=4
```

This result is consistent with the example in the Table 6.4.

In addition to these operators, Python has other operators, such as location operators and identity operators. Here is not an example; interested readers can refer to https://www.cnblogs.com/mehome/p/9439088.html.

6.2.1.7.2 Other Built-in Functions in Python   Only some commonly used built-in functions are given below (Table 6.5):

TABLE 6.5   Some Functions in Python

Some built-in Functions	Explanation	Examples
abs(x)	It is absolute value if real numb; it is modulus if complex numb.	abs(-1.1)=1.1
bin(x)	x must be an integer. Return a binary string that starts with 0b.	bin(5)='0b101'
chr(i)	Returns the str. format of characters whose Unicode code is i.	chr(97)='a'
ord(c)	Inverse func. of chr (), c is a single Unicode str., its int. is returned.	ord('a')=97
round(numb[, ndigits])	Rounding function, ndigits keep digits after decimal point.	round(2.34,1)=2.3
input([prompt])	Numb. or str. can be input; if there is no restriction, output is str.	Refer to below code
pow(base,x)	Equivalent to base**x, i.e. base^x	pow(2,3)=8
range(star,stop,step)	Actually, it is an immutable sequence type.	Refer to "Syntax"

Run the following code on Spyder to illustrate it:

```
print('abs(-1.1)=',abs(-1.1))
print('bin(5)=',bin(5))
print('chr(97)=',chr(97))
print('ord("a")=',ord('a'))
print('round(2.34,1)=',round(2.34,1))# this round function and Excel's ROUND function
no big difference
print('pow(2,3)=',pow(2,3))
numb=int(input('Please enter a number'))
print('numb=',numb)
print('numb's type is',type(numb))# here type is the type function
s=input('Please enter a string')
print('s=',s)
print('The type of s is',type(s))print('s=',s)
```

The result will be:

```
abs(-1.1)= 1.1
bin(5)= 0b101
chr(97)= a
ord("a")= 97
round(2.34,1)= 2.3
pow(2,3)= 8
Please enter a number 23
numb= 23
The type of numb is <class 'int'>
Enter a string abc
s= abc
The type of s is <class 'str'>
```

6.2.1.8 Custom Functions

While Python's built-in functions are very useful, they are not enough for the widespread application of computers. In practical applications, people need to create custom functions, which are essentially like "sub-programs" in Basic or Fortran. The main purpose of custom functions is to facilitate reuse. The syntax structure of custom functions is as follows: *def* function name (each relevant parameter): function body, *return* value. And the call format: variable name=function name (). This is too abstractable; it is better to give an example:

Example 6.1

Define a custom-function *Solve _ q(a, b, c)* to solve the quadratic equation $ax^2+bx+c=0$. Here, *Solve _ q* is the name of the custom-function, and a, b, and c are its formal parameters. You need to substitute actual numbers when calling this function. Based on the quadratic equation's solution formula, you can write the following Python code and run it in Spyder:

```
def Solve_q(a,b,c):#Custom-function about solving quadratic equations
    B=b*b-4*a*c # This is the discriminant of the quadratic equation
    if(B>0):#This if statement will be explained in detail in the "syntax" below
        BB=B**0.5#Square the discriminant B
        x1=(BB-b)/(2*a)#which is a real root of the equation
```

```
        x2=-(BB + b)/(2*a)#that is, the other real root of the equation
        return print('Two real roots: x1=','{:.3f}'.format(x1),',x2=','
        {:.3f}'.format(x2))
    if(B==0):
        x1=x2=-b/(2*a)
        return print('Two double roots, x1=x2=','{:.2f}'.format(x1))
    if(B<0):
        BB=(-B)**0.5
        y1=-b/(2*a); y2=BB
        x1=complex(y1,y2)# which is a complex root of the equation
        x2=complex(y1,-y2)#that is, the other complex root of the equation
        return print('two complex roots:x1=','{0.real:.2f}{0.imag:+.3f}
        j'.format(x1),',x2=', '{0.real:.2f} {0.imag:+.3f}j'.format(x2))
a=float(input('Please enter the first non-zero coefficient of the quadratic equation
a='))# add this float to ensure that the number entered is a floating-point number
b=float(input('Please enter the second coefficient of the quadratic equation b='))
c=float(input('Please enter the third coefficient of the quadratic equation c='))
#a,b,c=map(float,input('Please enter the three coefficients of the quadratic
equation a,b,c').split())
# The above 3 sentences can also be replaced by this one.
Solve_q(a,b,c)#solve the quadratic equation with coefficients a,b,c respectively
```

The result will be:

```
1. enter the first non-zero coefficient of the quadratic equation a=1.6
Enter the second coefficient of the quadratic equation b = 5.7
Enter the third coefficient of the quadratic equation c = 2.8
Two real roots: x1= -0.588,x2= -2.974
2. Enter the first non-zero coefficient of the quadratic equation a = 16
Enter the second coefficient of the quadratic equation b = -24
Enter the third coefficient of the quadratic equation c = 9
Two double roots, x1= x2= 0.75
3. Enter the first non-zero coefficient of the quadratic equation a = 2.5
Enter the second coefficient of the quadratic equation b = 3.2
Enter the third coefficient of the quadratic equation c = 1.3
Two complex roots: x1= -0.64+1.661j,x2= -0.64-1.661j
```

This shows that this custom-function is very good, but of course to really use it well, we need some effort. For example, the output format such as '{:.2f}'.format(x1) means print x1 with 2 decimal places [a simpler format output is '%.2f'%x1], while '{0.real:.2f} {0.imag:+.3f}j'.format(x1) indicates that the real and imaginary parts of the complex number x1 are printed with 2 and 3 decimal places, respectively. And return followed by what statement or parameter is also to be determined on a case-by-case basis; there is no standard statement.

6.2.2 Python's Syntax

First, it's important to note that Python is different from other computer languages in that it is an interpreted language, meaning it compiles and executes code line by line. The advantage is that it stops immediately when encountering an issue, but the downside is that code execution can be relatively slower compared to other languages. Due to this characteristic, the order of statements is crucial. For instance, as mentioned earlier, custom functions must be defined before they are called; otherwise, they cannot be invoked. One of the advantages of using Spyder is that it helps us check for syntax errors in the code before running it and provides useful prompts until the errors are fixed. Of course, this is a necessary condition for code execution, but for it to run correctly, it must also be tested under various simple scenarios. However, this is just a necessary condition, and it's essential to verify whether the actual results match expectations, especially in complex situations, where extra caution is required.

Next, let's discuss some of the syntax features in Python that differentiate it from other computer languages:

6.2.2.1 Indent

If a statement is followed by a colon, the statements controlled by it must be indented. One significant feature in Python is "indentation", which means (at least) indenting by four spaces to represent what would be brackets in other languages. For example, consider one statement from the custom-function mentioned earlier:

```
if(B==0):
    x1=x2=-b/(2*a)
    return print('Two double roots, x1=x2=','{:.2f}'.format(x1))
```

This represents that the *if* statement contains the following two statements, and the indentation is necessary; otherwise, the meaning would be entirely different. Fortunately, in Spyder, if a statement is followed by a colon, the indentation is automatically completed when you press Enter. If you feel that you no longer need indentation, you can move the cursor to the appropriate position.

6.2.2.2 Comment Statement

Comment statements have been encountered earlier. In Python, statements starting with "#" are comment statements. Similarly, in the custom-function code mentioned above, there are many comment statements, such as:

```
a=float(input('Please enter the first coefficient of the quadratic equation a='))# Add
this float to ensure that the number entered is a floating number
```

Here, "*# Add this float ensures that the input is a floating number*" is a comment statement. It is not an executable statement; its sole purpose is to describe the role of the statement. Sometimes, for explaining more complex code, several lines of comments may be needed. More often than not, when debugging a program, you might temporarily need to "skip" certain lines of code, and you can use triple quotes (single or double quotes are both acceptable) to turn them into non-executable statements. For example:

```
"""
D={'a':1,'b':2,'c':3,'d':4,'e':5}
print(list(D.keys()))#Print the list of keywords in dictionary D
print(list(D.values()))#Print a list of the values in the dictionary D
print(list(D.items()))#Print a list of key pairs in the dictionary D
print(D)#Print out the dictionary D
"""
```

In this pair of triple quotes, the five lines of code will not be executed.

6.2.2.3 If Statement

If statement, also known as a conditional statement, is indispensable in computer languages, as you have seen in the above example. There is a slight difference with other computer languages. Its syntax structure is that:

> *if* condition ：
>> code block
> [*elif* condition ：
>> code block]
> *else*:
>> code block

You can rewrite the three *if* statements of the custom-function in Example 6.1 into the above form:

```
if(B>0):
    BB=B**0.5# will prescribe the discriminant B.
    ......
elif(B==0):
    x1=x2=-b/(2*a)
    return print ('two double roots, x1=x2=','{:.2f}'.format(x1))
else:
    BB=(-B)**0.5
......
```

The effect is the same as the original code.

6.2.2.4 for Loop Statement

People who have learned other computer languages are familiar with the *for* loop statement. However, the *for* statement in Python has its own special features compared to the *for* statements in other computer languages. The syntactic structure of for is that it loops through the values in a list:

for **cyclic variable in list or range(star, stop, step):** #where star, step' default values are 0 and 1 respectively.

For example, range(positive integer N), it means: star=0, stop=N (same as slice only to N-1), step=1. *for* can be used to loop through the elements in a list, such as [2,5,7,8,13], which is probably the biggest difference from other computer languages. Look at the code running on Spyder:

```
for i in range(5):
    print('i=',i,',', ', ',end='')#"end=''" means no line break
print()# is equivalent to a line feed
s=0
A=[2,5,7,8,13]
for i in A:
    s+=i# is equivalent to s=s+i, that is, used to sum
    print('i=',i,',', ', ',end='')#"end=''" means no newline
print()# acts as a line feed
print('Sum of elements in list A=',s)
```

The result of the run is that:

```
i= 0, i= 1, i= 2, i= 3, i= 4,
i= 2, i= 5, i= 7, i= 8, i= 13,
The sum of the elements of the list A = 35
```

A list can also be assigned a value using a for loop, by running the following code on Spyder:

```
B=[i for i in range(5)]
print('B=',B)
C=[i for i in B]# is equivalent to C=B
print('C=',C)
```

The result of the run is that:

```
B= [0, 1, 2, 3, 4]
C= [0, 1, 2, 3, 4]
```

6.2.2.5 While *Loop Statement*

For non-professionals in computer science, using *for* loop statements is more common than using loop statements that start with *while*. The issue lies in not understanding how to use the *while* statement. In fact, in many cases, using the while statement is more appropriate than using the *for* statement. To explain this, let's first introduce the syntax structure of *while*:

while **condition :**
　　code block
　　[*else:*
　　　　code block]

Essentially, *while* and *for* statements are both loop statements. The main difference between the two is that the *for* statement theoretically needs to iterate through all elements in a range or a list, *while* stops the loop when a condition is not met (in a sense, it's equivalent to *for* with *if*), making it more flexible. Of course, you can achieve the same requirement by adding an *if* to a *for* loop.

Example 6.2

Write a simple code to determine if a natural number (not very large) is prime.

Solution

We can easily obtain the following Python code:

```
N=int(input('Please enter a natural number'))
a=N//2#'//' means integer division
while a>1:
  if N%a==0:
        print(N,'The maximum factor of 'is',a)
        break#that is, stop the loop
  a-=1# is equivalent to a=a-1
  else:print(N,'is a prime number')
```

The result will be:

```
Enter a natural number 49
The largest factor of 49 is 7
Enter a natural number 17
17 is a prime number
```

It would be possible to use a *for* statement, but it doesn't seem as "elegant" as the above code:

```
N=int(input('Please enter a natural number'))
N1=int(N**0.5) #because the maximum factor of N cannot exceed the root number N
for i in range(2,N1+1):#why add 1? because to ensure that i loops to (N1+1)-1=N1
      if N%i==0:
            print(N,'The maximum factor of 'is',max(i,N//i))
            break#that is, stop the loop
      if(i==N1-1):print(N,'is a prime number')
```

The result will be:

```
Enter a natural number 81
The largest factor of 81 is 27
Enter a natural number 17
17 is a prime number
```

From this, you can see that there are many ways to complete the same task. Sometimes, it's hard to say which method is best, and it depends on the specific problem. During the learning process, you should ask yourself questions. For example, if it's not a prime number, can you find all of its prime factors? This is indeed a challenging question, and below is a reference answer:

```
N=int(input('Please enter a natural number ='))
F=[]#Place the prime factors in an empty list
def Factor(N):
    N1=int(N**0.5)#because the maximum factor of N cannot exceed the root N
    if N1==1 and N==1:return print('Run result:')
    B=True#Boolean variable used only to solve the last prime factor
    for i in range(2,N1+1):
    if N%i==0:
        B=False#If N is the last prime factor, then B cannot be "False"
        N2=N//i
        F.append(i)
        if i==N2: #Here to cope with the case of the same prime factor
            F.append(i)
            return Factor(N2//i) # recursion is used here
        return Factor(N2) # recursion is used here
    if B == True:F.append(N)
Factor(N)#Run this custom-function to find the prime factor
if len(F)==0:print(N,'Neither prime nor composite')
elif len(F)==1:# list only one prime factor is itself a prime number
    print(N,'is a prime number')
else:print('The prime factor of N is',F)
```

The result will be:

```
Please input a natural number 1
1 is neither a prime number nor a composite number

Please enter a natural number 3
3 is a prime number

Please enter a natural number r 123
The prime factor of 123 is [3, 41].

Please enter a natural number 127
127 is a prime number

Please enter a natural number 120
The prime factor of 120 is [2, 2, 2, 3, 5].
```

For readers not familiar with C or Java languages, this "small program" has a highlight, namely, the use of "recursion". Recursion means that a "subroutine" calls itself, which is not allowed in Basic or Fortran languages, but it is allowed in Python, making the code concise and elegant. Of course, you can achieve the same without using recursion; if you're interested, you can try it yourself.

6.2.2.6 break, continue, pass Statements

The so-called *break* statement is the break loop statement, which has actually been used in the above example. For example, the following code is used when talking about the *while* statement:

```
while a>1:
    if N%a==0:
        print(N,'s maximum factor is',a)
        break# that is, stop the loop
    ........
```

That is, once the condition of N%a==0 is met, the execution of the "*print*" statement terminates the if statement but still continues to execute the *while* statement. Sometimes, this *break* statement is also used when debugging a program.

The *continue* statement "skips" the loop, not terminates it. Here is an example:

```
i=0
while i<50:
    i+=1
    if i==3:
        continue
    elif i==5:
        break
    print('i=',i,',',',',end='')
```

The result will be that:

```
i= 1,i= 2,i= 4
```

As you can see, in i==3, use *continue* to "skip" the 3, and in i=5, use *break* to terminate the while loop.

The *pass* statement is a "do nothing" statement that does nothing, so what is it used to do? It simply "occupies a place", just like the "white space" in a painting, and the "blank space" in an article is in this sense indispensable. If you run a piece of code on Spyder:

```
N=int(input('Please enter a natural number'))
if N<=10:
    pass #If you enter a number less than 10, then do nothing
else: print('This is a number greater than 10')
```

The result will be:

```
Please enter a natural number 7
Please enter a natural number 15
This is a number greater than 10
```

Is this code boring? Indeed, it is. However, it's essential to note that we can't only teach content that will be "useful later". It's impossible to predict exactly what you will need in the future, just as you can't stop eating after "the last bite". During this learning process, we will do our best to consider the needs of readers. The goal is not to become a professional in computer science but to become proficient in using Python in your field. Therefore, more specialized knowledge in Python, such as "encapsulation, inheritance, polymorphism, generators, iterators, exception handling", and more, will not be covered (Lutz, 2011, 2015). What we've discussed so far in terms of "morphology" and "syntax" is sufficient to address the needs of "intelligent fatigue statistics". Of course, in the sections below, we will introduce a few commonly used software libraries or "packages", especially those related to plotting, to enable more versatile applications.

6.3 INTRODUCTION TO SEVERAL IMPORTANT PYTHON MODULES (LIBRARIES OR PACKAGES)

6.3.1 Introduction to NumPy, Pandas, and Scipy

The advantages of Python are not only its simplicity and ease of learning but also its ability to easily utilize existing software libraries, making it a powerful choice across various application domains. For example, while Fortran was once considered the dominant language in scientific computing, Python has shown itself to be quite competitive. Likewise, there are numerous probability and statistics software packages like SAS, SPSS, MATLAB, and many others, but Python can encompass some of the best statistical and graphing software, making it incredibly versatile. Furthermore, Python is indispensable in the field of machine learning. Now, let's begin by introducing some important modules (often referred to as libraries or packages).

6.3.1.1 NumPy

According to online sources, "NumPy (Numeric Python) provides many advanced numerical programming tools, such as: matrix data types, vector processing, and sophisticated arithmetic libraries. It is designed to perform rigorous numerical processing....... can be used to handle tasks that would otherwise be done using C++, Fortran, or MATLAB, etc". [8] This package focuses on the concepts of matrices and multidimensional arrays. We believe you have some knowledge of matrices, which are important because this mathematical form can express many realistic contents. For example, in "Structural Mechanics", we often encounter "strain matrix", "stress matrix", "elasticity coefficient matrix", etc. In linear algebra, the matrix method is used to solve the solution of linear algebra, and furthermore, the least squares method can be solved to obtain the regression line, etc. What is a matrix? Simply put, it is a two-dimensional array. A list in Python is also a matrix, i.e., a list where each element is a list of the same length. Since Python can also create matrices, why do we need NumPy (in the future, it will be written as **numpy** in practical use) matrices? This is because numpy matrices are more "professional" and can perform various matrix operations directly. In Python, the matrix constructed from a list of lists is more "amateurish" and very inconvenient to use. This is the main reason to introduce the numpy package. The question is how to introduce the numpy package? Just write the following statement in front of the code:

import numpy as np # "import" means leading-in, and "as np" means that you only need to write np.

when using functions in numpy in the future. Note that in the code, it cannot be written as NumPy but as numpy, so in the future, it should be written as numpy regardless of the situation.

An example of how to create a matrix in numpy and how to perform the main matrix operations is given on Spyder:

```
import numpy as np
a=np.mat('1,2;3,4')#define the matrix directly in numpy using the string
b=np.mat('1,3;2,4')#Note that the lines are separated by semicolons and the elements in the
lines are separated by commas
print('matrix a=',a,'\n','matrix b=',b)#'\n' means line feed
c=np.mat(np.array(1,5).reshape(2,2))# here the reshape function turns a one-dimensional
array into a 2-dimensional matrix
d=c.T# indicates the transpose of matrix c
print('matrix c=',c,'\n','matrix d=',d)
```

The result will be:

```
matrix a= [[1 2]
 [3 4]]
 matrix b= [[1 3]
 [2 4]]
 matrix c= [[1 2]
 [3 4]]
 matrix d= [[1 3]]
 [2 4]]
```

As seen, both a and b as well as c and d have the same values. They achieve the same results using different methods, and the choice between them depends on the specific requirements.

Here, it is also important to point out the difference between arrays and matrices. From a mathematical point of view, a matrix is just a two-dimensional array, and there is no big difference between the two. But from the numpy point of view, the difference between the two is still relatively large, mainly because they have different data structures. This is because matrices are always two-dimensional and their structure is fixed, but arrays do not have a fixed number of dimensions, so the data structure is also variable. More importantly, matrices have operations that arrays cannot have, such as there is no "inverse operation" for arrays, but there is an inverse operation for matrices (square matrix), i.e., the inverse matrix. Look at the following code:

```
import numpy as np
a=np.mat('1,2;3,4')
b=np.array([[1,2],[3,4]])# Note that if it is a two-dimensional array must add more than
one square bracket
print('matrix a=',a,'\n','array b=',b)
aT=a.I #Seek the inverse matrix of matrix a
print('inverse matrix of matrix a a.T=',aT)
bT=b.I #Seeking the inverse of an array is meaningless and will definitely report an error
```

The result will be:

```
matrix a= [[1 2]
 [3 4]]
array b= [[1 2]
 [3 4]]
inverse matrix a of matrix a.T= [[-2. 1.]
 [ 1.5 -0.5]]
......
AttributeError: 'numpy.ndarray' object has no attribute 'I'# for array ".I" this operation
does not exist
```

This shows that matrix a and array b are the same in form, but matrix *a* can find its inverse matrix, but for array B, the so-called "inverse array" b.T is meaningless and will naturally report an error.

Now, we will talk about the matrix multiplication directly with "*", but in the array, "*" means that the corresponding elements multiply together, and the multiplication with matrix meaning is to use "dot", that is "dot product" in vectors. Let's look at the following code in Spyder:

```
import numpy as np
a=np.mat('1,2;3,4')#define the matrix directly in numpy using the string
b=np.mat('3,4;1,2')#a,b are matrices
c=np.array([[1,2],[3,4]])#c,d are arrays
d=np.array([[3,4],[1,2]])
print('matrix a*b=',a*b,'\n','array c*d=',c*d)
print('dot product of arrays c and d =',c.dot(d))
```

The result will be:

```
matrix a*b= [[ 5 8]
 [13 20]]
The array c*d= [[3 8]
 [3 8]]
```

```
The dot product of the arrays c and d c.dot(d) = [[ 5 8]
[13 20]]
```

It follows that the matrix's a*b ≠ is the array's c*d, but the matrix's a*b = the array's c.dot(d).

The last thing to mention is that the biggest difference between the array in numpy and the matrix in numpy is that the dimension of the array changes when doing the averaging operation, but the matrix always stays as two-dimensional. Here is the code that runs on Spyder:

```
import numpy as np
a=np.mat('1,2;3,4')
b=np.array([[1,2],[3,4]])
aM=a.mean(1)# parameter 1 means average by row, if 0 is taken then average by column, if
vacant then average over all elements
bM=b.mean(1)# parameter 1 means average by rows
print('Matrix a averaged by rows=',aM,'\n','Array b averaged by rows=',bM)
print('Matrix a-aM=',a-aM,'\n','Array b-bM=',b-bM)
```

The result will be:

```
Matrix a averaged by rows = [[1.5]
 [3.5]]
Array b averaged by rows = [1.5 3.5]
Matrix a-aM= [[-0.5 0.5]
 [-0.5 0.5]]
Array b-bM= [[-0.5 -1.5]
 [ 1.5 0.5]]
```

It can be seen that the average value obtained by averaging the matrix by rows is still a two-dimensional matrix, but the result of averaging the two-dimension array by rows has become a one-dimensional array. Therefore, the results of taking the average of a and b are different even in form. As for the numerous scientific calculation functions and statistical functions in numpy, we won't introduce them one by one here. You can check them online when you need them.

6.3.1.2 Pandas

In accordance with online sources, "Pandas is a NumPy based tool created to solve data analysis tasks. Pandas incorporates a large number of libraries and standard data models, providing the tools needed to efficiently manipulate large datasets. Pandas provides a large number of functions and methods that enable us to process data quickly and conveniently".[9] Here, it can be noted that pandas is a further development of numpy, the biggest development being the data structure. In the introduction to numpy, matrices and arrays are introduced in particular, and while they are widely used, the limitations of matrices cannot be ignored as the range of applications expands. The biggest limitation is that the elements in a matrix or array are required to be of the same type. Elements of different types cannot be handled by matrices. The simplest example is a table with not only data but also "headers" or "attributes" (or "labels") of the data (columns or rows). For this reason, a new data structure DateFrame has been introduced in pandas, whose elements in each column are likely to be of different types. Let's illustrate this with an example run in Spyder:

```
import pandas as pd #import pandas package
test=[];L1=range(11)
for i in L1:
        for j in L1:
                    dist=i+2*j
                    test.append((i,j,dist))
df=pd.DataFrame(test,columns=['A','B','Dist'])# create a data frame out of yourself
print('df (the first 3 lines of this data frame):\n', df.head(3))# print df the first 3 lines
of this data frame
wdf=df[(df['Dist']>=3)]# form a new data box
wdfL=list(wdf['A'])# will be a slice of the frame list
print('wdf(the first 3 lines of this data frame):\n',wdf.head(3))#print wdf the first
3 lines of this data frame print('wdfL=',wdfL[:12])#print wdL the first 11 elements of
this data frame list
ord=list(set(wdf['A'])) # will be a frame wdf 'A' column slice first into a set (in order
to remove duplicate elements) and then list
print('ord=',ord)
```

The result will be:

```
df (the first 3 lines of this data frame):
  A B Dist
0 0 0 0
1 0 1 2
2 0 2 4
wdf (the first 3 lines of this data box):
  A B Dist
2 0 2 4
3 0 3 6
4 0 4 8
wdfL= [0, 0, 0, 0, 0, 0, 0, 0, 0, 0, 1, 1, 1]
ord= [0, 1, 2, 3, 4, 5, 6, 7, 8, 9, 10]
```

It's important to note that the first column of the DataFrame serves as an "index column" and does not necessarily need to be numeric; it can also use text labels. Additionally, another significant role of Pandas is data reading. In practical situations, you often need to read external data, especially data stored in formats like Excel (though not limited to Excel). The data is typically structured as DataFrames. Here's a simple example of reading specific data from a fatigue test using Pandas, assuming the Excel table is located in the same folder as the code:

```
import pandas as pd
df = pd.read_excel('fatigueNum1.xlsx') # now Excel table with py code in a folder
print('df (the first 3 lines of this data box):\n',df.head(3))# print df the first 3 lines
of this data box
```

The running result is that:

```
Ord. Life Reliability Log life Standard fraction
0    1    124    90.909091  2.093422   -1.33517
1    2    134    81.818182  2.127105   -0.90845
2    3    135    72.727273  2.130334   -0.60458
```

Therefore, the data in this DataFrame can be referred to as an array as needed and can also be transformed into a matrix. Various graphs can be drawn as needed. And "fatigueNum1" just is Table 5.1 in Section 5.2.2.

6.3.1.3 Scipy

Scipy is, in fact, "based on NumPy, providing methods (function libraries) to compute results directly, encapsulating some higher-order abstractions and physical models. Let's say do a Fourier transform". In short, "a convenient, easy-to-use Python toolkit designed for science and engineering. It includes statistics, optimization, integration, linear algebra modules, Fourier transforms, signal and image processing, ordinary differential equation solvers, and more. It can basically replace MATLAB".[10] For example, if you want to calculate a definite integral, of course, numerical methods can be used to solve it using a computer, but after all, it is troublesome. In Scipy, you only need to give the integrand and the upper and lower limits of integration to get the result immediately. Suppose the following definite integral is required, $S = \int_{-1}^{1} \left(1 - x^2\right)^{1/2} dx$, by replacing $x = \sin y$, it can be obtained that $S = 2\int_{0}^{\pi/2} \cos^2 y \, dy = \pi / 2$ with the variable $x = \sin y$, but on Spyder, we can use the following code:

```
from scipy import integrate #import definite integral function from scipy
def g(x): #define the integral function
        return (1-x**2)**0.5
pi_2,err = integrate.quad(g,-1,1) #Integration result and error
print('pi=',pi_2*2,', computational error=',err) # integration result is half of π
```

The result will be:

```
pi= 3.141592653589797, calculation error= 1.0002356720661965e-09
```

In turn, this is a way to find π. Naturally, scipy for definite integrals is only one of the functions, the function of scipy.stas in probability and statistical calculations is precisely the foundation of the Zhentong Gao method that will be introduced in Chapter 8. Without this feature, the method cannot be achieved. We will introduce the details in Section 8.1.3.

There may be thousands of packages (libraries) available for different uses of Python, so we can only introduce a few that we think are more commonly used here, and we will encounter other packages as needed, so we can use them as we learn. The following is a very important package for Python, which will definitely be used.

6.3.2 Introduction to Matplotlib and its Key Points of Graphing

Matplotlib according to its official website is "a Python 2D plotting library that generates publication-quality graphics in multiple hardcopy formats and in a cross-platform interactive environment. Matplotlib is available for Python scripts, Python and IPython Shells, Jupyter notebooks, web application servers, and four GUI toolkits".[11] In a sense, it can be said that all the drawings that Excel can draw in Matplotlib can be easily drawn, and the former can also draw the drawings that cannot be drawn in Excel. For example, it is quite troublesome to draw more than two curves on the same graph in Excel, but this is just a piece of cake for Matplotlib. The most important thing is that Matplotlib can be used to draw the results of Python code directly without "human help", which is exactly what "intelligent fatigue statistics" has to do (Figure 6.6).

Matplotlib is so powerful that it is impossible to give a comprehensive introduction here, but only a "quick start", even if the reader can use it immediately, some details, and the so-called more advanced use of the

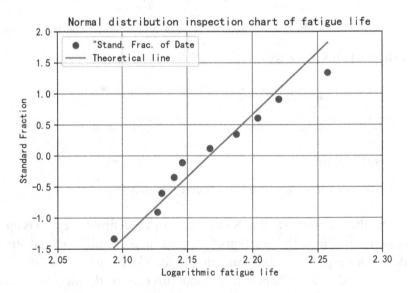

FIGURE 6.6 Normal distribution test chart of fatigue life.

reader can only slowly experience and improve in the process of use. Some of the most basic functions are listed below[12]:

```
import matplotlib.pyplot as plt # This library is necessary to import in order to draw
plots13
plt.rcParams['axes.unicode_minus'] = False# Solve the problem of displaying negative signs
on axes
plt.title('Time') # draw the "theme" of this graph
plt.title('E-N curve (E0='+str(E0)+', m='+mm+', C='+CC+')')# in the title to add variable
parameters (such as mm in this sentence) must be turned into a string (with the str function)
and then linked with '+'!
plt.xlim(0,1.0)#give the range of the x-axis
plt.ylim(0,1.0) #give the range of the y-axis
plt.xlabel("x")#set the horizontal axis label
plt.ylabel("y")#set the vertical axis label
plt.plot(x,y, "b^-",ms="6",label="mylabel")#You can also set the thickness of the line and
other styles according to your needs
plt.scatter(xx,yy,c='r',marker='o') # give a scatter plot, 'r' means red (red)
plt.scatter(x,y,c='b',marker='^') #give scatter plot, 'b' means blue(blue), '^' means Δ symbol.
plt.semilogx(XN,SY,'o') #give a semi-log scatter plot, as long as no color is specified
plt.semilogx(X2,YN,label='Fitting Curve')#draw a semi-logarithmic fit to the S-N curve
plt.grid(c='g')#Automatically generates green grid lines
plt.legend()#display the legend, while its position is automatically determined by the computer
plt.show# indicates that the graph is complete
```

Here is an example of directly plotting after reading the data in Excel (i.e., the fatigue life in column 3 of Table 5.2-1), the Python code is that:

```
import matplotlib.pyplot as plt#import the matplotlib plotting package
import pandas as pd#import the pandas package
df = pd.read_excel('fatigueNum1.xlsx')#Now the Excel table is in a folder with the py code
x = df['log life'].values# import the value of the independent variable x, here is the log
fatigue life
y=df['standard fraction'].values#import value of variable y, here is standard fraction
value of log fatigue life
y1=df['theoretical fraction'].values# import the value of another variable y1, here is the
theoretical fraction value of log fatigue life, if the distribution is normal
# Here is the code for graphing
plt.rcParams['font.sans-serif']=['SimHei']#Solve the Chinese display problem
plt.rcParams['axes.unicode_minus'] = False# Solve the problem of displaying the negative sign
of the coordinate axes
plt.title('Fatigue life normal distribution test chart') # draw the "theme" of this chart
plt.xlabel("log fatigue life")#set the horizontal axis label
plt.ylabel("standard fraction")#set the vertical axis label
plt.xlim(2.05,2.3)#gives the range of the x-axis
plt.ylim(-1.5,2.0) #give the range of the y-axis
plt.scatter(x,y,c='r',marker='o') #give the scatter plot, 'r' means red (red)
plt.plot(x,y1,label="theoretical straight line")
plt.grid()# automatically generate grid lines
plt.legend()# indicates that no specific position is specified
plt.show()# indicates the end of plotting
```

The running result is the Figure 6.6.

Isn't it beautiful? The key is to do it in one go. It can be compared to Figure 5.1.

6.4 AN EXAMPLE: NUMERICAL SOLUTION OF A TRANSCENDENTAL EQUATION

In Section 3.2, while studying the Weibull distribution, we encountered equation (3.31), which can be rewritten as the transcendental equation below:

$$\Gamma(1+1/x)^x = \ln2 \qquad\qquad (6.1)$$

How can we solve this equation? It's evident that there is no analytical solution; it can only be solved numerically. The most commonly used method for finding numerical solutions to equations is the Newton-Raphson method. The principle behind this method is relatively simple and was taught in calculus. For a continuous function, if the values at the endpoints of an interval have opposite signs, there must be at least one root within that interval. So, how do we find that root? The simplest approach is the "bisection method". You take the midpoint of the interval, evaluate the function at that point, and if the function has the same sign as one of the endpoints, you replace that endpoint with the midpoint, effectively creating a new interval that still contains at least one root and has half the length of the original interval. You repeat this process until you achieve the desired accuracy for the root. This is the algorithm behind the bisection method. Without a computer, this process can be very cumbersome. However, with a computer, it becomes straightforward.

Let's use Python to find a solution for equation (6.1). First, we'll transform equation (6.1) into a functional form:

$$f(x) = \Gamma(1+1/x)^x - \ln2 \qquad\qquad (6.2)$$

Example 6.3

It's clear that $f(x)$ is continuous on the interval $(0,\infty)$. We'll first plot $f(x)$ using a computer to get a visual understanding of the bisection method (Figure 6.7). We'll take the interval [1, 10] and use the following code:

```
import numpy as np # import the numpy "module" into
import math # because there is a gamma function in the math module
import matplotlib.pyplot as plt#import matplotlib plotting package
A=np.log(2);B=math.gamma(1);E=0.001;b1=1;b2=10#initial conditions
def F(B):# calculate F=A-np.power(math.gamma(1+1/B),B)
    return A-np.power(math.gamma(1+1/B),B)
if F(b1)*F(b2)>0: print('this solve is wrong,please try choice b1 or b2')
x=np.linspace(1,b2+1,100)# independent variable x first take [1,10] this interval,
is consistent with the two endpoints of the function value is different sign
y=[F(i) for i in x]# use the function F(x) to assign a value to y
y1=[0 for i in x]#draw a "y=0" line
plt.title(' graph of the f(x)=ln2-np.power(math.gamma(1+1/x),x) ')
plt.xlabel("value of independent variable x")# set the horizontal axis label
plt.ylabel("f(x)")# set the vertical axis label
plt.xlim(1,10)#give the range of the x-axis
plt.ylim(-0.3,0.1) #give the range of the y-axis
plt.grid()
plt.legend
plt.plot(x,y)
plt.plot(x,y1,linewidth=3,color='r')#linewidth=represents width
plt.show
```

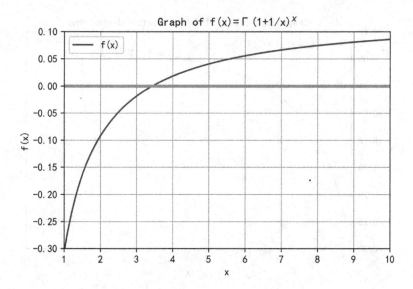

FIGURE 6.7   Example 6.3 schematic diagram.

Running this code will generate the Figure 6.7.

Example 6.4

With a slight modification of the previous code, we can obtain the zero value of the function f(x) using the bisection method (Figure 6.8):

```
import numpy as np # the numpy "module" into the
import math # because there is a gamma function in the math module
import matplotlib.pyplot as plt#import matplotlib plotting package
A=np.log(2);B=math.gamma(1);E=1.0e-5;b1=1;b2=10#initial conditions
def F(B):# calculate F=A-np.power(math.gamma(1+1/B),B)
return A-np.power(math.gamma(1+1/B),B)
if F(b1)*F(b2)>0: print('this solve is wrong,please try choice b1 or b2')
x=np.linspace(1,b2+1,100)# independent variable x first take [1,10] this interval,
is consistent with the two endpoints of the function value is different sign
y=[F(i) for i in x]# use the function F(x) to assign a value to y
Y=[0 for i in x]# draw a "y=0" line
x1=[]
def Solve(B1,B2):
    BI=(B2+B1)/2
    k=0#k represents the number of pairwise divisions
    while abs(F(BI))>E:
            BI=(B2+B1)/2#This is equivalent to recursion
            x1.append(BI)
            if F(B1)*F(BI)>0:#This is the core part of Newton's dichotomy
                B1=BI
            else: B2=BI
            k+=1
    return k,BI
k,BI=Solve(b1,b2)
print('k(number of dichotomy)=',k,',E(precision)=',E,',b={0:.5f}'.format(BI))
y1=[F(i) for i in x1]#assign a value to f(x) for the "valid" coordinate point
plt.title(' graph of the f(x)=ln2-np.power(math.gamma(1+1/x),x) ')
plt.xlabel("x")
```

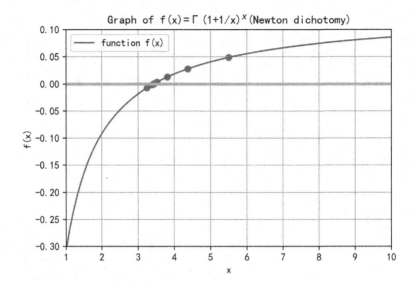

FIGURE 6.8   Example 6.4 schematic diagram.

```
plt.ylabel("f(x)")# set the vertical axis label
plt.xlim(1,10)#give the range of the x-axis
plt.ylim(-0.3,0.1) #give the range of the y-axis
plt.grid()#add grid lines
plt.plot(x,y,label='function f(x)')#draw a plot of f(x)
plt.legend()#Display the legend
plt.plot(x,Y,linewidth=3,color='y')#linewidth=represents width
plt.scatter(x1,y1,c='r',marker='o') #give the scatter plot of z approaching zero
one by one, 'r' means red (red)
plt.show
```

Running this code will provide the following result and the Figure 6.8:

```
k(number of pairs of points)= 14,E(accuracy)= 1e-05,b=3.43951
```

Example 6.5

In addition to using Newton's dichotomy can also use the iterative method to solve this transcendental equation. Considering that the derivative containing the gamma function is more challenging, we will use the "fixed-point method" here, which involves solving equations of the form $x = f(x)$. The equation (6.1) can be rewritten as:

$$b = \ln(\ln(2))/\ln(\Gamma(1+1/b)) = f(b) \tag{6.3}$$

Here's the Python code for this approach:

```
import numpy as np # import the numpy "module"
import math #because there is a gamma function in the math module
import matplotlib.pyplot as plt#import matplotlib plotting package
A=np.log(np.log(2))
def F(B):# calculate F=A/np.log(math.gamma(1+1/b)). To calculate the immovable
point equation b=F(b)
        return A/np.log(math.gamma(1+1/B))
b=[];e=1.0e-5# indicates the allowable error, which is adjustable
b0=9.2#initial value, can be taken randomly or can be specified
b.append(b0)
b1=F(b0);b.append(b1)
i=1
while abs(b1-b[i-1])>e:
        b1=F(b[i]);b.append(b1)#Use while statement to complete the iterative process
        i+=1
print('Initial value b0=',b0,",iterations=",i,',precision=',e,", b=",'%.5f'%b[i])
x1=[];y=[];k=0
for j in range(2*i+1):#To draw the iterative process
        if j%2==0:y.append(F(b[k]));x1.append(b[k]);k+=1
        else:y.append(b[k]);x1.append(b[k])
BB=str('%.5f'%b[i])#In order to show it in the title of the graph
plt.title('Solving the transcendental equation by (fixed point) iteration, b='+BB)
plt.xlim(0,10);plt.ylim(0,10)
plt.xlabel('b');plt.yLabel('f(b)')
plt.plot(x1,y,c='r',label='{bi}');
xx=[1.15+i*0.1 for i in range(10)];xx1=[2.5+i for i in range(8)]#To make the curve
a bit rounded
xx=xx+xx1;yy=[F(xx[i]) for i in range(len(xx))]#Note that at this point b cannot
take the value 1
plt.plot(xx,yy,c='b',label='F(b)')
plt.plot(xx,xx,c='k',label='y=x')
plt.legend(loc='best')
plt.show
```

Running this code will provide the following resultand and the Figure 6.9:

```
Initial value b0= 9.2, iterations= 20,prec.= 1e-05, b= 3.43955
```

From this, we can see that although the two algorithms are different, the results are the same. The Newton's method is relatively easy to understand and can be used for equations in any form. However, the fixed-point method requires the equation to be written in the form of $x=f(x)$, and not all equations can be written in this form. However, if the fixed-point method can be used, it tends to have a faster iteration speed (Figure 6.9).

So far, we have provided a simple introduction to two major "intelligent" tools: Excel and Python. However, our application of these tools has only just begun. In the third part of this book, we will start applying them to "fatigue statistics", making "fatigue statistics" more "intelligent". This is the main purpose of this book.

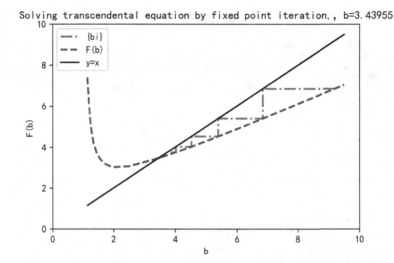

Solving transcendental equation by fixed point iteration, , b=3. 43955

FIGURE 6.9 Example 6.5 schematic diagram.

NOTES

1 Refer to https://zh.wikipedia.org/wiki/Python
2 Refer to https://www.cnblogs.com/rourou1/p/6039108.html
3 Refer to https://baike.baidu.com/item/%E4%B8%96%E7%95%8C%E7%BC%96%E7%A8%8B%E8%AF%AD%E8%A8%80%E6%8E%92%E8%A1%8C%E6%A6%9C/1600318 n
4 Refer to https://zhuanlan.zhihu.com/p/454028612
5 Refer to https://blog.csdn.net/Fristm/article/details/125654054?ops_request_misc=%257B%2522request% 255Fid%2522%253A%2522168983519716800 184132674%2522%252C%2522scm%2522%253A%252220140713.13010233 4..%2522%257D&request_id=168983519716800184132674&biz_id=0&utm_medium=distribute.pc_search_result.none-task-blog-2~all~top_positive~default-1-125654054-null-null.142^v90^control_2,239^v2^ insert_chatgpt&utm_term=%E7%BC%96%E7%A8%8B%E8%AF%AD%E8%A8%80%E6%8E%92%E8%A1%8C%E6%A6%9C&spm=1018.2226.3001.4187
6 Refer to https://www.jianshu.com/p/5601dab5c9e5
7 https://www.cnblogs.com/tangjiao/p/10369465.html
8 Refer to https://baike.baidu.com/item/numpy/5678437?fr=aladdin
9 Refer to https://baike.baidu.com/item/pandas/17209606?fr=aladdin
10 Refer to, https://www.jianshu.com/p/32cb09d84487
11 Refer to https://www.matplotlib.org.cn/intro/
12 Refer to https://blog.csdn.net/helunqu2017/article/details/78659490
13 Refer to https://blog.csdn.net/hulinku/article/details/79076449.

REFERENCES

Liu YX, Jiang Q and Kong ZQ. (2018), *Basic Python Tutorial*, Posts & Telecom Press, Beijing.

Lutz M (2011), *Learning Python*, 4th, Machinery Industry Press, Beijing.

Lutz M (2015), *Python Pocket Reference*, 5th, China Electric Power Press, Beijing.

III

Some Applications of Intelligent Fatigue Statistics

CHAPTER 7

Estimation of Population Parameters

7.1 POINT ESTIMATION OF POPULATION PARAMETERS

7.1.1 Mathematical Expectation and Variance of the Subsample Mean

Since the mean of the sample $\tilde{X}$ is a function of a random variable, i.e., it is itself a random variable:

$$\tilde{X} = (X_1, X_2,..., X_n)/n \qquad (7.1)$$

$$E(\tilde{X}) = E[(X_1, X_2,..., X_n)/n] = E(X_1, X_2,..., X_n)/n$$

By the independence of samples, we have that:

$$E(\tilde{X}) = [E(X_1) + E(X_2) + ... + E(X_n)]/n \qquad (7.2)$$

According to the central limit theorem, in the case of a large sample (when n is large):

$$E(\tilde{X}) \approx \mu \qquad (7.3)$$

Now, let's study the variance of $\tilde{X}$. Similarly, by definition:

$$Var(\tilde{X}) = Var[(X_1, X_2,..., X_n)/n] = Var[(X_1, X_2,..., X_n)/n^2]$$

DOI: 10.1201/9781003488477-10

As the samples are independent, and by noting (2.14):

$$\text{Var}(\tilde{X}) = [\text{Var}(X_1) + \text{Var}(X_2) + \ldots + \text{Var}(X_n)]/n^2 \tag{7.4}$$

Again, by the central limit theorem, we obtain:

$$\text{Var}(\tilde{X}) = \sigma^2/n \tag{7.5}$$

7.1.2 Point Estimation of the Population Mean and Variance

Generally, when $n > 50$, it is considered a large sample. However, in many cases, providing a large sample is not feasible either economically or in terms of time. In most cases, small sample sizes are the reality. Below, we will provide a method for point estimation of population parameters in the case of a small sample. In general, to estimate population parameters, we need to satisfy two requirements: consistency and unbiasedness. Unbiasedness was explained in Section 4.3.1. "Consistency" refers to the fact that "as the sample size increases, the value of the estimator gets closer and closer to the estimated population parameter".[1]

For example, the sample mean $\tilde{X}$ as an unbiased estimator of the population mean $\tilde{\mu}$ is evident:

$$E(\tilde{\mu}^\wedge) = \mu \tag{7.5}$$

where "$\wedge$" means "estimated value". But notice that from equation (7.2), it can be obtained that:

$$\tilde{\mu} = \tilde{X} \tag{7.7}$$

Now, assuming the fatigue life observation values for different specimens are N_1, N_2,..., N_n, the mean logarithmic fatigue life is given by:

$$\tilde{X} = (1/n) \sum_{i=1}^{n} \lg N_i \tag{7.8}$$

and,

$$\hat{\mu} = \tilde{X} = \lg[(\prod_{i=1}^{n} N_i)^{1/n}] \tag{7.9}$$

For a normal distribution, based on equation (3.13), it can be obtained that:

$$\mu = x_{50} = \lg N_{50} \tag{7.10}$$

Using equations (7.9) and (7.10), it can be obtained that:

$$\lg \tilde{N}_{50} = (1/n) \sum_{i=1}^{n} \lg N_i \rightarrow \tilde{N}_{50} = \left(\prod_{i=1}^{n} N_i \right)^{1/n} \qquad (7.11)$$

If the fatigue life of specimens follows the Weibull distribution, then its life can also be estimated as follows:

$$\hat{\mu} = \tilde{N} = (1/n) \sum_{i=1}^{n} N_i \qquad (7.12)$$

For an unbiased estimate of the population variance $(\hat{\sigma})^2$, it should satisfy the following equation:

$$E\left[(\hat{\sigma})^2 \right] = \sigma^2 \qquad (7.13)$$

Noting that the sample variance from (4.2) is:

$$s^2 = \left[1/(n-1) \right] \left[\sum_{i=1}^{n} X_i^2 - n(\tilde{X})^2 \right] \qquad (7.14)$$

and,

$$E(s^2) = \left[1/(n-1) \right] \left\{ E\left[\sum_{i=1}^{n} X_i^2 \right] - E\left[n(\tilde{X})^2 \right] \right\} \qquad (7.15)$$

where,

$$E\left[\sum_{i=1}^{n} X_i^2 \right] = n \left\{ E\left[\sum_i X_i^2 / n \right] \right\} = nE(X^2)$$

therefore:

$$E(s^2) = \left[n/(n-1) \right] \left\{ E\left[X^2 \right] - E\left[\tilde{X}^2 \right] \right\} \qquad (7.16)$$

However, from (1.23), it's known that:

$$E(X^2) = \text{Var}(X) + [E(X)]^2 \text{ and } E(X^2) = \sigma^2 + \mu^2 \qquad (7.17)$$

Replacing X with $\tilde{X}$:

$$E\left[\left(\tilde{X}\right)^2\right] = \text{Var}\left(\tilde{X}\right) + E\left(\tilde{X}\right)^2 = \sigma^2/n + \mu^2 \qquad (7.18)$$

So, from equation (7.16), we have:

$$E(s^2) = [n/(n-1)](\sigma^2 - \sigma^2/n) = \sigma^2 \qquad (7.19)$$

Therefore,

$$s^2 = \left(\hat{\sigma}\right)^2 \qquad (7.20)$$

The above discussion only covers the case of a single sample. If two samples are drawn from the population, with sizes n_1 and n_2, means $\tilde{X}_1$ and $\tilde{X}_2$, variances s_1^2 and s_2^2, respectively. Then, the estimated values for the population mean and variance are given by:

$$\hat{\mu} = \left(n_1 \tilde{X}_1 + n_2 \tilde{X}_2\right)/(n_1 + n_2) \qquad (7.21)$$

and,

$$\left(\hat{\sigma}\right)^2 = \left[(n_1 - 1)s_1^2 + (n_2 - 1)s_2^2\right]/(n_2 + n_2 - 2) \qquad (7.22)$$

First prove equation (7.21), i.e.:

$$E(\hat{\mu}) = E\left[\left(n_1 \tilde{X}_1 + n_2 \tilde{X}_2\right)/(n_1 + n_2)\right]$$

and,

$$E(\hat{\mu}) = \left[n_1/(n_1 + n_2)\right]E(X_1) + \left[n_2/(n_1 + n_2)\right]E(X_2)$$
$$= p_1 E(X_1) + p_2 E(X_2) \qquad (7.23)$$

This is the weighted average of the mathematical expectations of the two random variables. Since both samples come from the same population, their expectation (note that it is not the mean) should be the same as the population expectation $E(X_1) = E(X_2) = E(X) = \mu$, i.e.: $E(\hat{\mu}) = \mu \rightarrow$ q.e.d.

Now, let us prove equation (7.22). In fact:

$$E\left[\left(\hat{\sigma}\right)^2\right] = E\left\{\left[(n_1 - 1)s_1^2 + (n_2 - 1)s_2^2\right]/(n_1 + n_2 - 2)\right\}$$

and,

$$E\left[\left(\hat{\sigma}\right)^2\right]=\left[(n_1-1)/(n_1+n_2-2)\right]E\left(s_1^2\right)+\left[(n_2-1)/(n_1+n_2-2)\right]E\left(s_2^2\right)$$

(7.24)

Since both samples come from the same population, their expected values for variances should also be the same. Also, considering t equation (7.19), we have that:

$$E\left(s_1^2\right)=E\left(s_2^2\right)=E\left(s^2\right)=\sigma^2.$$

Therefore,

$$E\left[\left(\hat{\sigma}\right)^2\right]=\left[(n_1-1)/(n_1+n_2-2)\right]\sigma^2+\left[(n_2-1)/(n_1+n_2-2)\right]\sigma^2$$

$$=\sigma^2 \to q.e.d.$$

7.2 PARAMETER ESTIMATION OF GAUSSIAN DISTRIBUTION

7.2.1 Analytical Method

Now, we will apply the knowledge we have learned to parameter estimation in the fatigue testing process. The simplest distribution is naturally a Gaussian distribution. However, it must be noted in Section 3.1.4 that the fatigue life of the structure is more in line with the Gaussian distribution after taking the logarithm. This is a last resort, and it is also a traditional method in fatigue statistics, which can be used as an example of parameter estimation for logarithmic Gaussian distribution.

If a sample is known to have the following fatigue life values, N_1, N_2,..., N_n, for convenience, its logarithmic fatigue life can be taken as the sample mean:

$$\tilde{X} = (1/n)\sum_{i=1}^{n} lgN_i = \hat{\mu}$$

(7.25)

And the sample standard deviation can be calculated as:

$$s = \left\{\left[\sum_{i=1}^{n}\left(lgN_i\right)^2 -(1/n)\left(\sum_{i=1}^{n} lgN_i\right)^2\right]\Big/(n-1)\right\}^{1/2} = \hat{\sigma}$$ (7.26)

Taking note of (3.9), and replacing μ and σ with $\hat{\mu}$ and $\hat{\sigma}$, respectively, we can estimate the percentile:

$$\hat{x}_p = \hat{\mu} + u_p\hat{\sigma} = \tilde{X} + u_p s \qquad (7.27)$$

From this, we can determine the safety life corresponding to any reliability level as:

$$\hat{x}_p = \lg\left(\hat{N}\right)_p \qquad (7.28)$$

Here is an example, still using the data in Gao (1986) but not with "hand calculations" but using Excel, on the one hand, to increase the efficiency, on the other hand, also increased the accuracy and very convenient (Table 7.1).

To use Excel, it is enough to input only the data in the first column. The data in the second column can be directly calculated by using the logarithmic function LOG10() with the base of 10 in Excel, but in actual operation, only one data needs to be calculated, and then a column of data can be obtained immediately by using the method of copy. As for the mean value and the mean square deviation, they can also be directly obtained by using the functions AVERAGE () and STAEV.S () of Excel, respectively, without manual calculation at all. The results obtained are completely consistent with those calculated in Gao (1986).

In this way, the calculated results can be used to estimate the reliability, such as the safety fatigue life estimate corresponding to p=0.999.

TABLE 7.1   Logarithmic Lifetime

Fatigue Life (kc)	Log of Fat. Life
124	2.093421685
134	2.127104798
135	2.130333768
138	2.139879086
140	2.146128036
147	2.167317335
154	2.187520721
160	2.204119983
166	2.220108088
181	2.257678575
Average=	2.167361208
STAEV.S=	0.049846267

TABLE 7.2   Life Estimation Corresponding to p=0.999

$u_p = $ NORM.S.INV(0.001)=	−3.09023
$X_{99.9} = \tilde{x} + u_p * s =$	2.01332
$N = $ POWER$(10, X_{99.9})=$	103.11567

However, the numerical value is $1 - 0.999 = 0.001$, so the "standard fraction" obtained from this method can be directly obtained from Excel without looking up the table. By using $\hat{x}_{99.9} = \tilde{X} + u_p s$, the corresponding estimated safety life with reliability of 0.999 can be deduced as $N_{99.9} = 10^{\hat{x}_{99.9}}$. These calculations can be obtained from Excel, which is very convenient, as shown in Table 7.2.

Just be aware of using the relevant formulas. For example, NORM.S.INV () should be used to calculate standard fraction. Of course, NORM.INV () can also be used, but there are two more parameters. The results obtained in this table are consistent with those calculated in Gao (1986).

The physical meaning of the final $N_{99.9} = 103$(kc) is that only one in a thousand individuals have a fatigue life of less than 103 kc.

From this, the logarithmic normal probability density of population estimation can also be deduced:

$$f(x) \approx [1/(0.05(2\pi)^{1/2})]\exp[-(x-2.167)^2/(2*(0.05)^2)] \qquad (7.29)$$

7.2.2 Graphing Method

The graphing method introduced in Gao & Xiong (2000) requires the use of "normal coordinate probability coordinate paper" on the graph, completely by "hand", although it seems simple, but the error with actual use of the process is actually relatively large; efficiency is also not high. Of course, the computer can be used to draw this "normal coordinate probability coordinate paper", but it is still relatively troublesome and inconvenient to use. It is better to change the idea, using the "standard fraction (3.9)" to graph, so that on the one hand, you can visually check whether the data are normally distributed, and on the other hand, you can also calculate the population parameters to be estimated (the mean and standard deviation) by fitting the line graph. This graphing method has been described in some detail in Section 5.2.2 and has been obtained in Figure 5.2-1. This will now be discussed in a little more depth. Using the same data, i.e., according to Table 5.1, the fitted curve of the "standard fraction" line, i.e., the trend line in Excel (the "......." in the figure below) "), the following Figure 7.1 can be drawn, and the reader is invited to note the difference with Figure 5.2-1.

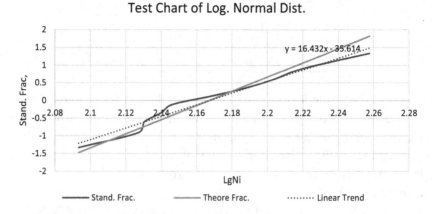

FIGURE 7.1   Normal distribution test chart.

Note that the fitted straight line of standard fraction is shown in Figure 7.1, that is, the line equation obtained by the least square method (this problem will be described in detail in Section 7.4) is, y=16.432x−35.614. Where 16.432 and −35.614 are the slope and intercept of the straight line, respectively. The meaning of x and y is:

$$\hat{x}_p = 16.432 * \lg(\hat{N}) - 35.614 \tag{7.30}$$

Furthermore, considering the definition of the standard fraction (3.9):

$$\hat{u}_p = \left(\lg(\hat{N}) - \hat{\mu}\right)/\hat{\sigma} \tag{7.31}$$

Therefore, when $\hat{x}_p = 0$, by (7.30), we have that: $\lg\hat{N}_{50} = 35.614/16.432 = 2.167$ Then, using (7.31), we have:

$$\hat{\mu} = \lg\hat{N}_{50} = 2.167 \tag{7.32}$$

When $\hat{x}_p = 1$, considering that NORM.S.DIST(1,TRUE)=0.84 in Excel, we can use (7.30) to find:

$$\lg(\hat{N})_{84} = (1 + 35.614)/16.432 = 2.228$$

But by (7.31), we have:

$$\hat{\sigma} = \lg(\hat{N})_{84} - \hat{\mu} = 2.228 - 2.167 = 0.0612 \tag{7.33}$$

Of course, you can also directly calculate the slope and intercept of the fitted line using Excel functions. In this case, you can use the "logarithm of fatigue life" (as the x-axis) and "standard score" (as the y-axis) columns from Table 5.1, along with the SLOPE and INTERCEPT functions (Table 7.3):

Therefore, the results of comparing the analytical method with the graphic method are listed as below (Table 7.4):

This shows that the results of the two methods are almost the same. This indicates that the population of this sample does conform to a logarithmic normal distribution, because the analytical method starts from the assumption that the population is normally distributed. The graphical method estimates the population mean and standard deviation from a fitted straight line of standard fractions corresponding to the reliability estimates. If the population distribution is not normal, the difference between the two will be significant. This will be observed in the next section "Estimation of the population parameters of the Weibull distribution".

Incidentally, the so-called "theoretical standard fraction" curve (as shown in Figure 7.1) is obtained using the analytical data, i.e., its equation has the same form as (7.31):

$$\hat{u}_p = \left(\lg\left(\hat{N}\right) - \hat{\mu}\right)/\hat{\sigma} \tag{7.34}$$

However, here, $\hat{\mu}$ and $\hat{\sigma}$ were taken from the analytical method's data as 2.167 and 0.05, respectively. So, based on equation (7.34), we have:

$$\hat{u}_p = 20.0 * \lg(\hat{N}) - 43.4 \tag{7.35}$$

The difference in slope and intercept between equations (7.30) and (7.35) reflects the significant difference in standard deviation.

TABLE 7.3 For Calculating Slope and Intercept with Excel

SLOP of fit. line of stand. frac.=	16.43185
INTERCEPT of fit. line of stand. Frac.=	−35.6138

TABLE 7.4 Comparing the Analytical Method with the Graphic Method

	$\hat{\mu}$	$\hat{\sigma}$
Anal. Method	2.167	0.05
Graf. Method	2.167	0.06

7.3 PARAMETER ESTIMATION OF WEIBULL DISTRIBUTION

7.3.1 Analytical Method

Similar to the Gaussian distribution, for the Weibull distribution, based on the previous results, we can have the following estimates for population parameters:

$$\tilde{N} = (1/n)\Sigma_i lgN_i = \hat{\mu} \tag{7.36}$$

and median value:

$$N_m = \hat{N}_{50} \tag{7.37}$$

as well as standard deviation:

$$s = \left\{\left[\Sigma_i (N_i)^2 - (1/n)(\Sigma_i N_i)^2\right]/(n-1)\right\}^{1/2} = \hat{\sigma} \tag{7.38}$$

Using $\hat{\mu}$ and $\hat{\sigma}$ instead of μ and σ, respectively, let's illustrate an example (Table 7.5).

According to equation (3.30a), we have:

$$557 = \tilde{N} = \hat{\mu} = (\hat{N}_a - \hat{N}_0)\Gamma(1+1/\hat{b}) + \hat{N}_0 \tag{7.39}$$

From Table 7.5, we also have that

$$\hat{N}_{50} = (540 + 550)/2 = N_m = 545$$

Also using equation (3.24):

$$\ln(2) = [(545 - \hat{N}_0)/\hat{N}_a - \hat{N}_0)]^b \tag{7.40}$$

And from equation (3.2-14a):

$$132.2^2 = s^2 = (\hat{N}_a - \hat{N}_0)^2[\Gamma(2/\hat{b}+1) - \Gamma^2(1/\hat{b}+1)] \tag{7.41}$$

TABLE 7.5 Date of Fatigue Life (kc)

350	380	400	430	450	470	480	500	520	540
550	570	600	610	630	650	670	730	770	840
$\tilde{N}$ =	557	s =	132.15						

The above three equations, (7.39), (7.40), and (7.41), have three unknowns: $\hat{N}_a$, $\hat{N}_0$ and b which are theoretically solvable. However, all three equations are transcendental equations, making it somewhat challenging to solve analytically. Using a computer to solve them is relatively easier. But before using a computer, we need to simplify them. We can use the elimination method, and for simplicity, let's set:

$$s = 132.2, \ N_{av} = 557, \ N_m = 545, \ L2 = \ln(2) \tag{7.42}$$

From equation (7.41), we can derive:

$$(\hat{N}_a - \hat{N}_0) = s/\{[\Gamma(2/\hat{b}+1) - \Gamma^2(1/\hat{b}+1)]\}^{1/2} \tag{7.43}$$

And from equation (7.39), we can derive:

$$\hat{N}_0 = N_{av} - \{s/[\Gamma(2/\hat{b}+1) - \Gamma^2(1/\hat{b}+1)]\}^{1/2}\Gamma(1+1/\hat{b}) \tag{7.44}$$

Substituting (7.43) and (7.44) into (7.40):

$$L2 = \{\{N_m - N_{av} + \{s/[\Gamma(2/\hat{b}+1) - \Gamma^2(1/\hat{b}+1)]\}^{1/2}\Gamma(1+1/\hat{b})\}/\{s/[\Gamma(2/\hat{b}+1) - \Gamma^2(1/\hat{b}+1)]\}^{1/2}\}^b$$

Simplifying further:

$$N_{av} - N_m + s[L2^{1/b} - \Gamma(1+1/\hat{b})]/\{[\Gamma(2/\hat{b}+1) - \Gamma^2(1/\hat{b}+1)]\}^{1/2} = 0 \tag{7.45}$$

Example 7.1

Solving this equation above is relatively easy, similar to (3.35) or using Newton's dichotomy to solve it. Run the following code on Spyder:

```
import numpy as np
import math
import matplotlib.pyplot as plt
N=[350,380,400,430,450,470,480,500,520,540,550,570,600,610,630,650,670,730,770,840]
s=np.std(N,ddof=1);Nav=np.average(N);Nm=np.median(N)
L2=np.log(2);e=1.0e-6;b1=1.0;b2=10.0;LS=len(N)
def F(b):
E=math.gamma(1+2/b)-np.power(math.gamma(1+1/b),2)
F=s*np.power(E,-0.5)
return Nav-Nm+F*(np.power(L2,1/b)- math.gamma(1+1/b))
if F(b1)*F(b2)>0: print('this solve is wrong,please try choice b1 or b2')

def Solve(B1,B2):
BI=(B2+B1)/2
k=0
while abs(F(BI))>E:
    BI=(B2+B1)/2#This is equivalent to recursion
```

```
        if F(B1)*F(BI)>0:
            B1=BI
        else: B2=BI
        k+=1
    return BI, k
b,k=Solve(b1,b2)
print('k(number of dichotomy)=',k,',E(precision)=',e,', b^={0:.5f}'.format(b))
E=math.gamma(1+2/b)-np.power(math.gamma(1+1/b),2)
F=s*np.power(E,-0.5)
N0=Nav-F*math.gamma(1+1/b)
Na=N0+F
print(' N0^={0:.0f}'.format(N0),', Na^={0:.0f}'.format(Na),'(kc)')
print(' Check whether the above calculation is correct as follows:')
Nav1=(Na-N0)*math.gamma(1+1/b)+N0#i.e.,(7.3-4)
S=(Na-N0)*np.power(E,0.5)#i.e.,(7.3-6)
D1=np.power((Nm-N0)/(Na-N0),b)#i.e.,(7.3-5)
print('Nav1=','%.0f'%Nav1,', S=','%.1f'%S,', ln(2)-D1=','%.2e'%(D-D1))
Na0=Na-N0
def f(i):
    return (b/Na0)*np.power((N[i]-N0)/Na0,b-1)*np.exp(-np.power((N[i]-N0)/Na0,b))
Y=[f(i) for i in range(LS)];YY=[0]+Y;NN=[N0]+N
y =stats.norm.pdf(NN, loc=Nav,scale=s)
N01=str('%.0f'%N0);b1=str('%.2f'%b);NA=str('%.0f'%Na)
plt.title('fitting curve by WD: b='+b1+',N0='+N01+',Na='+NA)
plt.ylim(0,0.0035)
plt.xlim(250,900)
plt.xticks()
plt.xlabel("N(kc)")
plt.ylabel("PDF")
plt.hist(N, bins=5, facecolor='b',density=True, alpha=0.3,label='fatigue data')
plt.plot(NN,YY,c='r',label='fit curve by WD')
plt.plot(NN,y,c='g',label='fit curve by GD')
plt.grid()
plt.legend(loc='lower left')
plt.show()
```

Running the results yields the Figure 7.2 and the following results:

```
k(number of dichotomy)=24,E(precision)=1e-06, b^=2.21062
N0^=280, Na^=593 (kc)
```

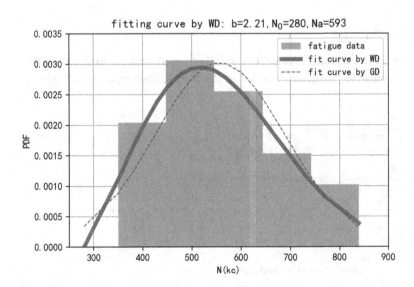

FIGURE 7.2   Weibull distribution fitting curve.

According to these parameters, the corresponding Weibull distribution PDF, i.e., equation (3.19), can be obtained as follows (Figure 7.2):

$$f(N) = (2.21/313)[(N-280)/313]^{1.21}\exp\{-[(N-280)/313]^{2.21}\} \quad (7.46)$$

You can also use the fixed-point method to solve equation (7.45). In this case, the equation becomes:

$$b = \ln D/\ln\{(N_m - N_{av})\{[\Gamma(2/b+1) - \Gamma^2(1/b+1)]\}^{1/2}/s + \Gamma(1+1/b)\} \quad (7.47)$$

The algorithm and code are similar to the example in Section 6.4 (readers may wish to do it), and the running result is,

```
initial value of b0= 9.2, the number of iterations=15, the precision=1e-06, and b=2.21062
N0^= 280, Na^= 593 (kc)
```

This is consistent with the result of Newton's dichotomy.

7.3.2 Graphing Method

As in the case of normal distribution, "Weibull probability graph paper" is not used here, but the line fitting method of Weibull distribution is used. The principle is as follows. We note that the reliability of the Weibull distribution (3.27) gives:

$$P = \exp\{-[(N_p - N_0)/(N_a - N_0)]^b\} \quad (7.48)$$

and,

$$\ln(1/P) = [(N_p - N_0)/(N_a - N_0)]^b$$

and

$$\ln[\ln(1/P)] = b\ln(N_p - N_0) - \ln(N_a - N_0) \quad (7.49)$$

Note that the left side of the above formula may be negative when the reliability is high. In order to make the graph look "prettier", both sides can be multiplied by −1. Additionally, the natural logarithm on the right can be changed to the logarithm with base 10. According to the base changing formula of logarithm:

$$\ln A = \lg A/\lg(e) = C\lg A$$

where,

$$C = 1/\lg(e) = 2.303 \tag{7.50}$$

(7.49) can become:

$$-\ln[\ln(1/P)] = -bC\lg(N_p - N_0) + bC\lg(N_a - N_0) \tag{7.51}$$

Let's set:

$$Y = kX + a \tag{7.52}$$

where,

$$Y = -\ln[\ln(1/P)], \ X = \lg(N_p - N_0), \ k = -bC, \ a = -k\lg(N_a - N_0)$$

So, we can derive:

$$b^\wedge = -k/C, \ \hat{N}_a - \hat{N}_0 = 10^{-a/k} \tag{7.53}$$

Using the data in Table 7.5, the following chart can be obtained in Excel (Figure 7.3):

Note that the data of Gao (1986) is used here, and $\hat{N}_0 = 300$. The obtained shape parameter $b = 1.775$ is also in good agreement with $b = 1.74$ of Gao (1986). However, it seems that it is far from the $b = 2.21$ obtained by the previous analytical method. What is the reason for this? The acceptable

ordinal(i)	Ni	P^	lg(Ni-N₀^)	-lnln(1/P^)
1	350	95.2381	1.69897	3.020227
2	380	90.47619	1.90309	2.301751
3	400	85.71429	2	1.869825
4	430	80.95238	2.113943	1.554433
5	450	76.19048	2.176091	1.302197
6	470	71.42857	2.230449	1.08924
7	480	66.66667	2.255273	0.90272
8	500	61.90476	2.30103	0.734859
9	520	57.14286	2.342423	0.580505
10	540	52.38095	2.380211	0.435985
11	550	47.61905	2.39794	0.29849
12	570	42.85714	2.431364	0.165703
13	600	38.09524	2.477121	0.035543
14	610	33.33333	2.491362	-0.09405
15	630	28.57143	2.518514	-0.22535
16	650	23.80952	2.544068	-0.36122
17	670	19.04762	2.568202	-0.50575
18	730	14.28571	2.633468	-0.66573
19	770	9.52381	2.672098	-0.855
20	840	4.761905	2.732394	-1.11334

Test Chart of Weibull Distribution

k=	-4.08794	b^=-k/C=	1.77505	C=1/lg(e)=	2.303
a=	10.10323	N_a^-N_0^=	296.1234	N_a^=	596.1234
r=	-0.99794	R²=	0.995875	N0^=	300

FIGURE 7.3　Test chart and table of Weibull distribution $\left(\hat{N}_0 = 300\right)$.

reason is that these data are not ideal Weibull distribution, so "fitting" from different angles will lead to inconsistent conclusions. To prove this idea, change $\hat{N}_0 = 300$ to $\hat{N}_0 = 280$ obtained by the analytical method and then look at the difference of the fitted straight line (Figure 7.4).

It is not difficult to find that the shape parameters have changed significantly, that is, $b = 2.0$ at this time, and most importantly, the correlation coefficient r has also increased, that is, from 0.997 to 0.999. In this sense, the analytical method should be better than the graphical method. In fact, just from the diagram, there is really no essential difference between the two. Only numbers can distinguish the subtle differences between the two. In this sense, we still need to speak with numbers, that is, we need to distinguish the advantages and disadvantages of fitting with numbers. This is also an important task of intelligent fatigue statistics.

Then, from (3.30a) and (3.2-14b), $\hat{\mu}$ and $\hat{\sigma}^2$ can be calculated, respectively.

If taking: $\hat{N}_a = 597.4$; $\hat{N}_0 = 280$; $\hat{b} = 2.0$, it can be obtained:

$(N_a{}^\wedge - N0^\wedge)\Gamma(1+1/b)+N0^\wedge = \mu^\wedge =$	**561.242**	$\hat{\sigma} =$ **146.837**
$\left(\hat{N}_a - \hat{N}_0\right)^2\left[\Gamma(2/b+1)-\Gamma^2(1/b+1)\right]=\left(\hat{\sigma}\right)^2 =$		**21,561.12**

This is not much different from the result of Gao (1986), which is a calculation error.

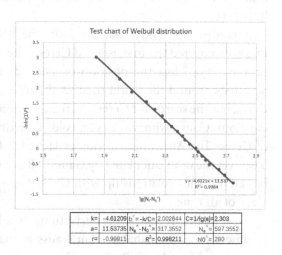

FIGURE 7.4 Test chart and table of Weibull distribution $\left(\hat{N}_0 = 280\right)$.

7.4 LEAST SQUARE METHOD

7.4.1 Principle of Least Square Method

The least square method has been known for a long time, but it is a very "painful" thing to really do by hand, and the calculation is too troublesome. Now, with computers, the calculation is not a problem; the problem is how to apply it. This requires a brief review of the meaning of this method. One phenomenon is worth noting, that is, in 1855, Darwin's cousin Galton found that the height of parents predicts children's height. A very interesting fact is that he also found that the children of tall parents are often not as tall as their parents, and the children of short parents are often taller than their parents, that is, the next generation has "regression" to the average height of the next generation.[2] This phenomenon also coincides with the saying that "good luck is hard to come by, and bad luck doesn't always follow". Or, in general, statistical distributions are overwhelmingly Gaussian distributions with large middle and small ends. Therefore, the arithmetic average of the statistics makes sense. However, we also find that two statistics with the same mean are still very different because they have different mean squared deviations. Therefore, mean squared deviation is also a very important concept. The so-called least square method was first started with linear fitting of data, which actually reflects regression. If the fit is successful, the linear relationship between the two types of data can be found. Of course, there are other functions (such as polynomials, exponential functions, and other primitive functions) fit; these will not go into the discussion, we here focus on the linear fit, the simplest and most important.

According to Wikipedia, "In 1801, the Italian astronomer Giuseppe Piazzi discovered the first asteroid Ceres. After 40 days of follow-up observations, Piazzi lost the position of Ceres as it orbited behind the Sun. Scientists around the world then used Piazzi's observations to search for Ceres, but the search for Ceres was fruitless according to most people's calculations. Gauss, who was 24 years old at the time, also calculated the orbit of Ceres. The Austrian astronomer Heinrich Oberth rediscovered Ceres based on the orbit calculated by Gauss".[3] It seems that least square method has an uncanny relationship with Gauss,[4] one of the greatest mathematicians of all time.

Now, we briefly review the derivation of the least square method, which is almost indistinguishable from Gauss at that time but of course more concise and "modern". Readers who are not interested in mathematical proofs can "skip" the process. However, the authors suggest the reader to "review" the mathematical knowledge through this proof.

Suppose there is a data set (x_i, y_i) $(i = 1, 2,..., n)$ and a trend line is to be found to express the direction indicated by this data set. For this purpose, it can be assumed that this trend line is:

$$Y = kX + d \tag{7.54}$$

The key is to find the slope k and the intercept d. The problem is that n is much greater than 2, so this equation cannot have an exact solution. We can only hope that the solution to this equation can be more reasonably accepted. To do this, we can assume:

$$e_i = y_i - (kx_i + d) \tag{7.55}$$

Notice that e_i can be positive or negative, so it's reasonable to measure the error by squaring it. We can define:

$$E = \sum_{i=1}^{n} e_i^2 = \sum_{i=1}^{n} (y_i - kx_i - d)^2 \tag{7.56}$$

It's quite evident that if we can minimize E, we can consider the resulting line acceptable:

$$\partial E / \partial k = 0 \rightarrow -2 \sum_{i=1}^{n} x_i (y_i - kx_i - d)$$

$$= 0 \rightarrow \sum_{i=1}^{n} x_i y_i - k \sum_{i=1}^{n} x_i^2 - d \sum_{i=1}^{n} x_i = 0$$

And,

$$\partial E / \partial d = 0 \rightarrow -2 \sum_{i=1}^{n} (y_i - kx_i - d) \rightarrow \sum_{i=1}^{n} y_i - k \sum_{i=1}^{n} x_i - nd = 0$$

thus,

$$k \sum_{i=1}^{n} x_i^2 + d \sum_{i=1}^{n} x_i = \sum_{i=1}^{n} y_i x_i \text{ and } k \sum_{i=1}^{n} x_i + nd = \sum_{i=1}^{n} y_i \tag{7.57}$$

For convenience, let's define the (arithmetic) mean values of X and Y as:

$$\tilde{X}=\Sigma_{i=1}^{n}x_i/n,\ \tilde{Y}=\Sigma_{i=1}^{n}y_i/n$$

Then, we can express k and d as:

$$k=\sum_{i=1}^{n}\left(x_iy_i-n\tilde{X}\tilde{Y}\right)\Big/\sum_{i=1}^{n}\left(x_i^2-n(\tilde{X})^2\right),\quad d=\tilde{Y}-k\tilde{X}$$

and,

$$k=\sum_{i=1}^{n}\left(x_i-\tilde{X}\right)\left(y_i-\tilde{Y}\right)\Big/\sum_{i=1}^{n}\left(x_i-\tilde{X}\right)^2,\quad d=\tilde{Y}-k\tilde{X}\qquad(7.58)$$

The theoretical derivation so far is very "beautiful", but the actual calculation according to this formula is not so elegant. This is probably one of the main reasons why many people do not like probability statistics and probably not many people like to repeat calculations. It is precisely the advent of computers that has freed people from such heavy computational work. It also made intelligent fatigue statistics possible.

The degree of correlation between two sets of data can of course be "fitted" by a straight line, but the degree of fit is related to the specific data. In order to describe this "correlation" in a more quantitative way, the concept of the so-called "correlation coefficient" is introduced. There are various definitions of the correlation coefficient, but the most used is the following definition proposed by the British statistician Pearson (Fisz, 1978):

$$r=E[(X-\mu_x)(Y-\mu_y)]/\{E[(X-\mu_x)^2]\{E[(Y-\mu_y)^2]\}^{1/2}=Cov(X,Y)/\sigma_x\sigma_y\qquad(7.59)$$

However, this definition is equivalent to the definition given in Gao & Xiong, (2000):

$$r=L_{XY}/(L_{YY}L_{XX})^{1/2}\qquad(7.60)$$

where,

$$Cov(X,Y)=L_{XY}/n=\sum_{i=1}^{n}y_ix_i-(1/n)\left(\sum_{i=1}^{n}y_i\right)\left(\sum_{i=1}^{n}x_i\right)=L_{YX}/n$$

$$(7.61)$$

and,

$$L_{YY}/n = Var(Y); \; L_{XX}/n = Var(X) \quad\quad (7.62)$$

It is easy to see that k in (7.58) can be written as: $k = Cov(X,Y)/Var(X)$. Again, if $Cov(X,Y) = 0$, then the slope of the line $k = 0$ and there will be no linear correlation between the two variables.

Based on the sign of k (positive, negative, or zero), we can categorize the correlation between two variables into three cases: positive correlation, negative correlation, and no correlation. Positive correlation means that the two variables change in the same direction, while negative correlation means they change in opposite directions. No correlation implies that the two variables have no relationship (independence).

Example 7.2

Data are from Gao (1986). The axial loading test of a notched specimen of a certain material shows that the stress concentration coefficient $K_t = 4$ and the average stress $S_m = 14 \; kgf/mm^2 = 140 \; Mpa$.[5] The stress amplitude and corresponding fatigue life are shown in the table below. The least square method is used to fit a straight line of lgS_a about lgN (Figure 7.5).

This gives the equation of the fitted line,

$$lgS_a = 2.811 - 0.258*lgN \quad\quad (7.63)$$

In Excel, the slope and intercept of the fitted line can be obtained directly from the least square method by using the two built-in functions SLOPE and INTERCEPT, and it is easy to draw a graph, which is undoubtedly very convenient. In Python, of course, there

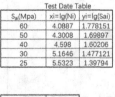

Test Date Table

S_a(Mpa)	xi=lg(Ni)	yi=lg(Sai)
60	4.0887	1.778151
50	4.3008	1.69897
40	4.598	1.60206
30	5.1646	1.477121
25	5.5323	1.39794

Slop: k=	-0.25754
Intercept:d=	2.810769
Correlation:r=	-0.99385

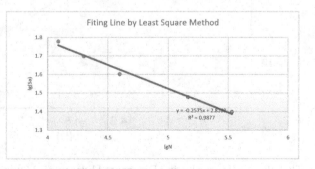

FIGURE 7.5   Chart and table of an example using LSM.

are many ways to fit a line by least squares, especially in connection with "machine learning", which has a great future. However, here we will only introduce two methods. One is the so-called "stupid method", which is to write a code in Python language according to the least square method; the second is to directly obtain the slope and intercept by calling the existing least squares function in scipy. Let's first take a look at the code that runs in Spyder using the "stupid method". Of course, there are various ways to write it, and readers can try it out themselves (Figure 7.6):

```python
import numpy as np
import matplotlib.pyplot as plt
A =np.mat([[4.0887,60], [4.3008,50], [4.598,40], [5.1646,30], [5.5323,25]])
(m,n)=np.shape(A)
x=[];y=[]
for k in range(m):
        x.append(A[k,0]);y.append(np.log10(A[k,1]))
meanX=np.mean(x);meanY=np.mean(y)
dx=x-meanX;dy=y-meanY
sumXY=0;sqX=0;sqY=0
for i in range(m):
        sumXY+=dx[i]*dy[i]
sqX+=dx[i]**2;sqY+=dy[i]**2
k=sumXY/sqX
d=meanY-k*meanX;r=sumXY/np.power(sqX*sqY,0.5)
print('Slope:k=%.3f'%k,',Intercept:d=%.3f'%d,',Correction:r%.3f'%r)
plt.scatter(x,y,c='blue',marker='o',s=60)
```

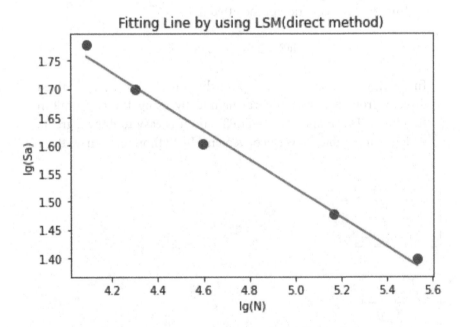

FIGURE 7.6   Linear Fitting Diagram of Example 7.2 Obtained Using LSM (Direct Method).

```
x.sort
y=[k*xi+d for xi in x]
plt.plot(x,y,'r')
plt.title('Fitting Line by using LSM(direct method)')
plt.xlabel('lg(N)')
plt.ylabel('lg(Sa)')
plt.show()
```

Running the results yields the Figure 7.6 and the following results:

```
Slope :k=-0.258,Intercept: d=2.811,Correction:r-0.994
```

Example 7.3

Now, we use the second method, directly calling the ready-made least squares function in scipy to obtain the slope and intercept. The only difference from the code above is how to obtain the slope and intercept. Therefore, the initial part, i.e., the data acquisition part and the plotting part, remains the same:

```
import numpy as np
from scipy.optimize import leastsq
import scipy.stats as stats
import matplotlib.pyplot as plt
A =np.mat([[4.0887,60], [4.3008,50], [4.598,40], [5.1646,30], [5.5323,25]])
(m,n)=np.shape(A)
x=[];y=[]
for k in range(m):
        x.append(A[k,0]);y.append(np.log10(A[k,1]))
X=np.array(x);Y=np.array(y)
def linearF(p):
        k,d=p
        return Y-k*X-d
result=leastsq(linearF,[1,1])
k,d=result[0]
r,f=stats.pearsonr(X,Y)
print("Slope: k=%.3f"%k,"Itercept: d=%.3f"%d,"Correction:r=%.3f"%r)
plt.scatter(x,y,c='blue',marker='o',s=60)
x.sort
y=[k*xi+d for xi in x]
plt.plot(x,y,'r')
plt.title('Fitting Line by LSM(using scipy)')
plt.xlabel('lg(N)')
plt.ylabel('lg(Sa)')
plt.show()
```

Running the results yields the Figure 7.7 and the following results:

```
Slope: k=-0.258 Intercept: d=2.811 correction: r=-0.994
```

And the previous results are almost identical (Figure 7.7).

7.4.2 Best Linear Fit of Fatigue Performance Data

In fact, many of the previous examples show that we have been using the least square method to fit fatigue life data for a long time. Now, we will

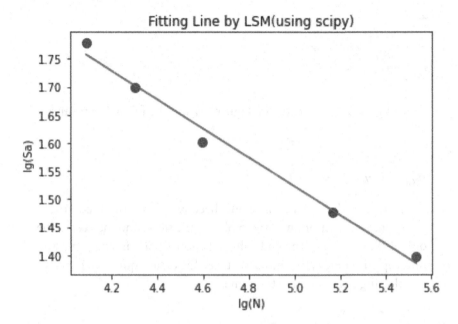

FIGURE 7.7 Linear Fitting Diagram of Example 7.2 Obtained Using LSM (Introducing Scipy).

specifically talk about the application of the least square method to the so-called S-N and P-S-N curves. It is important to note that the P-N curves discussed above are fatigue lives under the same stress, and it is obvious that the fatigue life of the same specimen under different stresses is not the same. The so-called S-N curve is the value of different stresses and fatigue life relationship curve. The so-called P-S-N curve is the S-N curve under the same conditions of reliability. These two curves reflect the fatigue properties of the material or component, called the "fatigue performance curve" (Gao, 1986) (Figure 7.8).

It should be noted that the S-N curve in the above figure is also called the "median" S-N curve, i.e., the median value with 50% reliability, i.e., μ_i for each S. The P-S-N curve is the x_p corresponding to the specified reliability P. The horizontal distance between the two curves is $\mu - x_p = |\mu_p \sigma|$. That is, each point of P-S-N can be understood as the displacement where $|\mu_p \sigma|$ occurs.

In order to explore the variation law of the fatigue performance curve people have done a lot of work and obtained the following three empirical formulas (Gao, 1986).

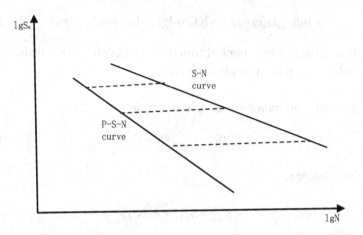

FIGURE 7.8   Curves of S-N and P-S-N.

7.4.2.1 Power-Law Expression

$$S_a^m N = C \qquad (7.64)$$

This formula represents the power-law relationship between stress amplitude S_a and life N under the stress ratio $R = S_{min}/S_{max}$ condition. The values of m and C depend on the specific material, component shape, and loading method, determined through experiments. Taking the logarithm of both sides of the equation:

$$m \lg S_a + \lg N = \lg C, \text{ i.e., } \lg N = \lg C - m \lg S_a \qquad (7.65)$$

This implies that the so-called power-law relationship corresponds to a linear relationship in logarithmic coordinates. Since stress amplitude and maximum stress are directly related, there exists the following relationship:

$$S_{max}^m N = C \qquad (7.66)$$

It is clear that the values of m and C in equations (7.64) and (7.66) are different under the same conditions, except for representing constants.

7.4.2.2 Exponential Function Expression

$$\exp(m S_{max}) N = C \qquad (7.67)$$

Take logarithms on both sides:

$$mS_{max}lg(e)+lgN=lgC \rightarrow lgN=lgC-mS_{max}lg(e) \qquad (7.68)$$

This shows that the exponential function relationship is equivalent to the linear relationship for a single logarithm.

7.4.2.3 Three-Parameter Power-law Expression

$$(S_{max}-S_0)^mN=C \qquad (7.69)$$

It can also become:

$$S_{max}=S_\infty(1+A/N^a) \qquad (7.70)$$

where S_0, m, C and S_∞, A, a are all constants and have the following relations:

$$S_0=S_\infty; \ m=1/a; \ C=(AS_0)^m \qquad (7.71)$$

It is easy to see that when $N\rightarrow\infty$, $S_{max} \rightarrow S_\infty=S_0$.

The power-law and exponential relationships are suitable for the middle-life range segments in S-N curves, while the three-parameter power-law expression is suitable for intermediate and long-life range segments because it is more flexible with three parameters.

Now, revisiting the example from Section 7.4.1, the fitted linear equation (7.63) obtained through the least square method can be transformed into:

$$lgN=2.811/0.258-(1/0.258) \ lgS_a \qquad (7.4-72)$$

and,

$$NS_a^{3.876}=10^{10.895}=7.852*10^{10}$$

Comparing this equation with (7.4-11), we can find that:

$$m=3.876, \ C=7.852*10^{10}$$

7.4.3 Linear Fitting of P-N Curve

As mentioned earlier, Gaussian probability paper can be used to intuitively check whether the fatigue life of a specimen satisfies the Gaussian distribution. However, due to the precision issues of Gaussian probability paper and the inconvenience compared to using a computer, Excel is now available, which can easily obtain other data by simply inputting fatigue life data. This method is especially convenient as it does not require table lookup to obtain "standard scores". Moreover, it uses the principles of least squares to draw a fitted line graph. This graph makes it easy to determine whether the fatigue life of the specimen fits the Gaussian distribution well and provides the values of correlation coefficients for judgment. Let's further discuss this method.

In fact, this method is based on the definition of standard fraction (3.9):

$$u_p = (x_p - \mu)/\sigma \tag{7.73}$$

Here, "standard scores" defined by the above equation are related to x_p, which is replaced by $\lg N_p$. By calculating the corresponding standard scores based on the corresponding reliability, it becomes the following linear equation:

$$\lg N_p = a + b u_p \tag{7.74}$$

Consequently, from the above two equations, we immediately obtain:

$$a = \mu, b = \sigma \tag{7.75}$$

From (7.4-21), we have:

$$\lg N_p = \mu + \sigma u_p \text{ and } X = \mu + \sigma Y \tag{7.76}$$

It is crucial to note that although the above equation is very similar to the linear fitting line $Y = kX + d$, the graphs in the coordinate system are different. However, their coefficients have the following relationship:

$$k = 1/\sigma, d = -\mu/\sigma \tag{7.77}$$

From this, it can be seen that μ and σ give physical meanings to the coefficients of the linear fitting line. The slope is inversely proportional to the variance, and the intercept is also inversely proportional to the variance but directly proportional to the mean. The reason for these results is essentially that the data distribution is approximate to the Gaussian distribution. If it were an ideal Gaussian distribution, the data points would perfectly match the fitted line. Now, let's use the data from Section 5.2 to create the following table and graph in Excel (Figure 7.9).

In fact, the so-called P-N diagram can also be drawn in Excel. That is, the reliability is taken as the vertical axis (Figure 7.10).

7.5 MAXIMUM LIKELIHOOD ESTIMATION

The renowned British statistician Ronald A. Fisher (1890–1962) proposed the practical "Maximum Likelihood Estimation"(MLE) method in the early 20th century based on Gaussian principles.[6] Essentially, it refers to determining the parameters of a distribution pattern by observing data when the distribution pattern is known. This is a form of "point

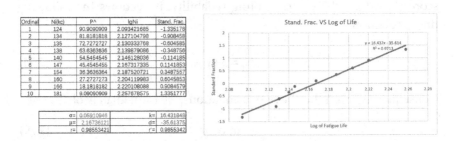

FIGURE 7.9   Chart and Table of Standard Fraction vs Log. Fatigue Life.

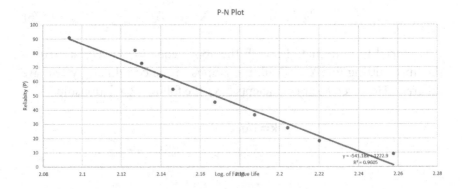

FIGURE 7.10   Chart of Reability vs Log. Fatigue Life.

estimation" for parameters. So, what sets it apart from other statistical methods? The key distinction lies in its use of calculus to find the maximum value, which is why it is called the "maximum likelihood" method. This description might seem somewhat abstract, so let's provide some examples to illustrate it.

7.5.1 The Discrete Case

Let's assume there is a very large container with only two types of balls: black and white. However, we don't know the proportion of these two types. To determine this proportion, or in other words, the probability p of picking a black ball, we need to perform experiments to estimate it. Without emptying the entire container, we cannot obtain an exact value for p. Therefore, conducting ball-picking experiments can only provide an estimate of p. The question is, which estimate is more likely?

For example, let's say we conducted 15 ball-picking experiments and obtained 4 black balls and 11 white balls. What is the most probable value for p? It is not difficult to calculate that the probability of picking a black ball, denoted as L, is $L = C_{15}^4 p^4 (1-p)^{11}$. What value of p maximizes L? By taking the derivative of L with respect to p and setting it equal to zero using calculus, we can find $p = 4/15$. This result is not surprising and aligns with the previously discussed frequency definition of probability. In this case, it has been formalized but doesn't seem to offer much novelty.

7.5.2 The Continuous Case

Let's use the Gaussian distribution as an example. Suppose we have n observations, $x_1, x_2, \ldots, x_n$ and assume they follow a Gaussian distribution $N(\mu, \sigma^2)$, and they are independent. In this case, we can define the likelihood function as follows (Gao, 1986):

$$L(\mu, \sigma^2 \mid x_1, x_2, \ldots, x_n) = f(\mu, \sigma^2 \mid x_1) f(\mu, \sigma^2 \mid x_2) \ldots f(\mu, \sigma^2 \mid x_n)$$

$$= \prod_{i=1}^{n} f(\mu, \sigma^2 \mid x_i)$$

$$= \left[1/(2\pi)^{1/2} \sigma\right]^n \prod_{i=1}^{n} \exp\left[-(x_i - \mu)^2 / 2\sigma^2\right]$$

$$= \left[1/(2\pi)^{1/2} \sigma\right]^n \exp\left\{-\sum_{i=1}^{n}\left[(x_i - \mu)^2 / 2\sigma^2\right]\right\}$$

$$\tag{7.78}$$

To maximize this likelihood function, we can take partial derivatives with respect to μ and σ^2 and set them to zero. However, before taking partial derivatives, we take the natural logarithm of equation (7.78) for ease of computation:

$$\ln L = -n\ln(2\pi)/2 - (n/2)\ln\sigma^2 - \left(1/2\sigma^2\right) \sum_{i=1}^{n} \left[(x_i - \mu)^2\right] \quad (7.79)$$

Therefore:

$$\partial(\ln L)/\partial\mu = 0 , \Sigma_{i=1}^{n}(x_i-\mu)=0 \text{ and } \hat{\mu}=(\Sigma_{i=1}^{n}x_i)/n = \tilde{X} \quad (7.80)$$

And,

$$\partial(\ln L)/\partial\left(\sigma^2\right)=0, \quad -n/2\sigma^2 +\left(1/2\sigma^4\right)\sum_{i=1}^{n}\left[(x_i -\mu)^2\right]=0$$

so,

$$\hat{\sigma}^2 = \sum_{i=1}^{n}\left[(x_i -\mu)^2\right]\Big/n \quad (7.81)$$

As seen, the estimation of the population mean is similar to the earlier discussed sampling estimation, but the estimation of variance is biased rather than unbiased. Unbiased estimation has a denominator of $(n-1)$, whereas biased estimation uses n. This implies that there is a significant difference between the two when the sample size is relatively small. So, can we say that unbiased estimation is better than biased estimation, or conversely, biased estimation is better than unbiased estimation? In reality, it depends on the specific situation. In cases with a limited number of samples, using the maximum likelihood principle, biased estimation is often more suitable. However, with repeated sampling, according to the central limit theorem, unbiased estimation is closer to the actual situation. Of course, when n is large, the differences between the two are not significant.

It's worth noting that in most general textbooks (Meyer, 1986), the application of the maximum likelihood method is usually limited to the Gaussian distribution. There is hardly any mention of other multi-parameter distributions such as the Weibull distribution, mainly because both the formula derivation and numerical calculations are more complex. In Section 8.3, we will discuss how to use the Zhentong Gao method and the MLE to solve the three-parameter estimation problem of the Weibull distribution.

NOTES

1 Refer to https://wenwen.sogou.com/z/q704504614.htm?ch=fromnewwen-wen.pc
2 Refer to http://www.gerenjianli.com/Mingren/19/iocdba25pg.html
3 Refer to https://zh.wikipedia.org/wiki/ least square method
4 Gauss, 1777–1855, a proud mathematician and physicist in Germany, whose Gaussian distribution is just one of his contributions.
5 1 pa = 1 Newton/1 m^2, 1 Mpa = 100,000 pa = 1 Newton/mm^2 ≈ 0.1 kg/cm^2. So, 1 kgf/mm^2 = 10 Mpa
6 https://en.wikipedia.org/wiki/Maximum_likelihood_estimation

REFERENCES

Fisz M (1978), *Wahrscheinlichkitsrechnung und Mathematische Stastistik*, Shanghai Science and Technology Press, Shanghai, p. 81.

Gao ZT (1986), *Fatigue Applied Statistics*, National Defense Industry Press, Beijing, pp. 132, 138, 142, 143, 147, 154.

Gao ZT & Xiong JJ (2000), *Fatigue Reliability*, Beihang University Press, Beijing, p. 121

Meyer PL (1986), *Introduction Probability and Statistics*, Higher Education Press, Beijing, p. 371.

CHAPTER 8

Zhentong Gao Method

8.1 ORIGIN OF THE ZHENTONG GAO METHOD

8.1.1 Problems with the Analytical Method

When researching the fatigue life of components, a crucial question arises: when encountering a set of actual life data, how do we determine whether it follows a Gaussian distribution, a Weibull distribution, or some other distribution. In other words, there is a problem of determining the so-called goodness of fit. Let's illustrate this with the data from Figure 7.9 in Section 7.4.3. Even if we take the data from Section 7.4.3 and transform it into logarithmic form for Gaussian distribution fitting, it doesn't fit very well when looking at the graph. What if we try to fit it directly with a Weibull distribution? To do this, we need to calculate the three parameters of the Weibull distribution as per the method in Section 7.3.1. By making a simple modification to the Python code from Section 7.3.1, we can obtain the following results:

```
N= [124, 134, 135, 138, 140, 147, 154, 160, 166, 181]
Nav= 147.9,s= 17.317, Nm= 143.5
k(number of dichotomy)= 26,E(precision)= 1e-07,estimated shape parameter b=1.221
Estimated safe life N0=126.86,Estimated characteristic life Na=149.32 (100kc)
The following is the verification: Nav1= 147.900, s1= 17.317, Nm1= 143.50
```

This result may look good at first glance, but there is a significant problem. The safe life N0=126.86 is actually larger than N[0]=124, which is clearly contradictory to the definition of safe life. This suggests that these data may not conform well to the Weibull distribution, and there may be some issues with the algorithm used to calculate the three parameters of the Weibull distribution. Without doubt, these data cannot be 100% in

DOI: 10.1201/9781003488477-11

line with the Weibull distribution, and in fact, they cannot be 100% in line with the Gaussian distribution either. So, which one fits better? To address this, improvements in the Weibull distribution calculation algorithm are needed, and a comparative standard should also be provided.

8.1.2 Fitting Criteria and the Coefficient of Determination

Improvements to the algorithm will be given in the next section, and a more reasonable comparison standard is given here. In Section 4.4, the expected value estimate of reliability equation (4.6) $\hat{P} = 1 - i/(n+1)$ is given. At the same time, it is particularly emphasized that this formula has nothing to do with the distribution of data, so it can be regarded as a "benchmark". That is, if the same set of data can be described by different distributions, the criterion for its quality is the correlation coefficient between the corresponding reliability and this benchmark. The one with a larger correlation coefficient is superior. Is this statement reasonable?

Let's conduct a more rigorous theoretical analysis here (Trivedi, 2015). Consider two random variables, X and Y, and we want to design a function $d(x)$ such that the random variable $d(X)$ approximates Y as closely as possible under certain conditions. When $d(X)$ is used to predict the value of Y, there is typically a modeling error ε, and the true value can be expressed as:

$$y = d(x) + \varepsilon \tag{8.1}$$

The random variable corresponding to ε is: $D = Y - d(X)$ (8.2)

The ideal function we seek is the one that minimizes $E[D^2]$, and this function is usually referred to as the least squares regression function (curve) of Y with respect to X. It can be seen that the simplest $d(x)$ is a linear function $a + bx$.

Now, let's assume that:

$$d(x) = a + bx \tag{8.3}$$

This simplifies the problem to finding appropriate values for a and b that minimize the following expression:

$$G(a, b) = E[D^2] = E[(Y - d(X))^2] = E[(Y - a - bX)^2] \tag{8.4}$$

This expression is essentially no different from equation (7.56), so the derivation in Section 7.4.1 is similar, with the only difference being in the notation used (μ_x and μ_y are the means of X and Y; $\sigma_x^2 = \text{Var}(X)$, $\sigma_y^2 = \text{Var}(Y)$). However, more conclusions can be deduced. It is evident from (7.58) and (7.59) that:

$$b = r(\sigma_y/\sigma_x), \ a = \mu_y - r\mu_x(\sigma_y/\sigma_x) \tag{8.5}$$

and (Trivedi, 2015),

$$E\left[D^2\right] = G(a,b) = G\left(\mu_y - r\mu_x\left(\sigma_y/\sigma_x\right), r\left(\sigma_y/\sigma_x\right)\right)$$

$$= \sigma_y^2 + \left[r\left(\sigma_y/\sigma_x\right)\right]^2 \sigma_x^2 - 2r\left(\sigma_y/\sigma_x\right)r\left(\sigma_x\sigma_y\right) \tag{8.6}$$

$$= \sigma_y^2 - r^2\sigma_y^2 = \left(1 - r^2\right)\sigma_y^2$$

Therefore,

$$\sigma_y^2 = r^2\sigma_y^2 + E\left[D^2\right] \tag{8.7}$$

Thus, $r^2\sigma_y{}^2$ can be understood as the variance of the linear part Y of Y to X, which is the "explainable" variance, while $E\left[D^2\right] = \left(1 - r^2\right)\sigma_y^2$ can be seen as the "residual variance". The ratio of the explained variance of Y with respect to X to the total variance is defined as the coefficient of determination (R^2), as follows:

$$R^2 = r^2\,\sigma_y^2/\sigma_y^2 = r^2 \tag{8.8}$$

This is the physical meaning of the coefficient of determination and its relationship with the correlation coefficient. That is, if the two variables are linearly related, the above formula is true, but if they are not linearly related, the above formula is not true. In nonlinear correlation, the coefficient of determination can also be used to judge the degree of fit between the fitting curve and the original curve. Therefore, the magnitude of the coefficient of determination can be used as a criterion for the quality of the fit.

In fact, it can be further demonstrated by (Trivedi, 2015):

$$R^2 = \Sigma\left(Y_i - \tilde{y}\right)^2 \Big/ \Sigma\left(y_i - \tilde{y}\right)^2 = SSR/SST \tag{8.9}$$

and,

$$R^2 = 1 - SSE/SST = 1 - Var(Y-y)/Var(y) \tag{8.9a}$$

Here, Y_i and y_i are on the fitted curve and the corresponding actual values, respectively, and both Y and y can be viewed as random variables. $\tilde{y}$ is the mean of the actual values. SSR (sum of square for regression)= $\Sigma_{i=1}^{n}\left(Y_i - \tilde{y}\right)^2$ is the sum of squared regression errors, SSE (sum of square for error)=$\Sigma_{i=1}^{n}\left(Y_i - y_i\right)^2$ is the sum of residual squares, and SST(sum of square for total)=$\Sigma_{i=1}^{n}\left(y_i - \tilde{y}\right)^2$ is the sum of total deviation squares. It is easy to prove that in the case of linear regression,

$$SST = SSE + SSR \tag{8.9b}$$

But in the case of nonlinear regression, the above equation does not hold and the coefficient of determination must be used (8.9a).

8.1.3 The Birth of Zhentong Gao Method

In the previous section, it was pointed out that there is a problem with calculating the three parameters of the Weibull distribution as introduced in Section 7.3.1. While it can be ensured that the mean, variance, and median of the calculated Weibull distribution match the observed values, it cannot guarantee that the calculated safe life will always be smaller than the minimum observed value. This naturally indicates that the observed values may not necessarily conform to the Weibull distribution 100%. On the other hand, it could also be a problem with the algorithm itself, meaning that it cannot guarantee the absence of this situation. So, is there an algorithm that can ensure that this situation does not occur? To address this problem, we can start with the two-parameter Weibull distribution.

Noting that (3.23), we have: $f(x) = (b/\lambda)(x/\lambda)^{b-1}\exp[-(x/\lambda)^b]$. Then, from (3.27a), we can find its reliability:

$$p = 1 - F(x_p/\lambda) = \exp[-(x_p/\lambda)^b] \tag{8.10}$$

Its estimated value can be obtained using (4.6), so after taking the natural logarithm of both sides, we get:

$$\ln(\ln(1/p_i)) = b\ln(x_i) - b\ln(\lambda) \qquad (8.11)$$

where, $p_i = 1 - i/(n+1)$. Let's define:

$$Y_i = \ln(\ln(1/p_i)), \; X_i = \ln(x_i) \qquad (8.12)$$

which gives us:

$$Y_i = bX_i + d \qquad (8.13)$$

where $d = -b\ln(\lambda)$, i.e.:

$$\lambda = \exp(-d/b) \qquad (8.14)$$

Then, b, d, and λ can be found by the least square method. This seems simple. Unfortunately, the premise of using Weibull distribution with two parameters is to set the location parameter, i.e., the safe life $x_0 = 0$. In actual situations, x_0 is not equal to zero, and it is precisely because x_0 is not equal to zero that it shows the superiority of Weibull distribution. Thus, the parameters b and λ and the corresponding correlation coefficient r are all functions of x_0. Obviously, x_0 that maximizes the correlation coefficient r can be obtained through analytical methods. However, the derivation by this method is still troublesome and error-prone. It is better to use Python and the scipy library to directly find out the x_0 corresponding to the maximum value of r in the interval $0 \leq x_0 < x_{min}$ (x_{min} here is the minimum value of the given data). Therefore, there is no need to derive and solve complex transcendental equations, but the required results can be obtained intelligently directly through Python. In theory, it seems to be divided into "two steps", but in the actual code, it is completed in one go. In fact, while finding the x_0 corresponding to the maximum value of r, b and λ are determined at the same time. This is the Zhentong Gao algorithm, which is based on the characteristics of Python, and is referred to as the Zhentong Gao method (abbreviated as ZT Gao method):

1. Input the original data, and if the original data is unordered, sort it first.

2. Scipy in Python can be used to directly traverse the interval $(0, x_{min})$ of possible values of x_0 with a given precision to find the x_0 that maximizes the correlation coefficient, i.e., x_{0max}.

3. Notice that the correlation coefficient in scipy is in fact calculated by first finding the corresponding coefficients b and d of the linear equation in least squares, $\lambda = \exp(-d/b)$. Thus, once x_{0max} is fixed, the corresponding parameters b and λ of Weibull distribution are also obtained almost simultaneously.

Having this ZT Gao effectively solves the problem of estimating the three parameters of Weibull distribution, which has become an important contribution of this book. The corresponding Python core code is that:

```
def fn(N0):
    x=[np.log(N[i]-N0) for i in range(LS)]
    y=[np.log(np.log(1/(1-i/(LS+1)))) for i in range(1,LS+1)]
    X=np.array(x);Y=np.array(y)
    return X,Y
def linearF(p,X,Y):
    b,d=p
    return Y-b*X-d
def maxF(N00):
    r=[];b=[];lamda=[]
    for i in N00:
        X,Y=fn(i)
        result=leastsq(linearF,[1,1],args=(X,Y))
        b1,d1=result[0]
        r1,f=stats.pearsonr(X,Y)
        r.append(r1);b.append(b1);lamda.append(np.exp(-d1/b1))
    return r,b,lamda
r,b,lamda=maxF(N00)
km=r.index(max(r))
bm=b[km];Lm=lamda[km];N0=N00[km];rm=r[km]
```

Readers who wish to master Python should take a closer look at these three custom-function, as they are the basis for implementing the ZT Gao method. Of course, this is not optimal code and can be fully optimized further, and the reader is expected to do so.

Example 8.1

Performing a trial calculation on the example above, we obtain the following results (Figure 8.1):

```
Nav=, 147.9,s= 17.317, Nm= 143.5
r= 0.986339,bm= 1.836,lamda= 40.33,N0= 113.15
```

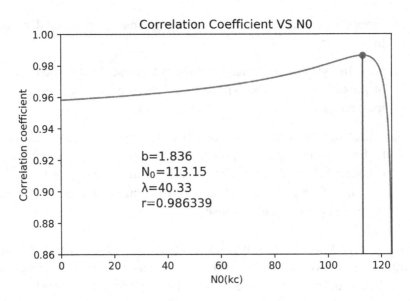

FIGURE 8.1   Chart of correlation coefficient VS N0.

According to this new algorithm, it can be determined that N0=113.15<124 is satisfied. However, the mean and variance at this time are different from the mean and variance of the original data. Who can guarantee that the mean and variance are the same as the population mean and variance in the case of a small sample? This code can be further improved to show the advantages and disadvantages of the data when viewed as Weibull distribution and Gaussian distribution, respectively. The running results are as follows:

```
Between Gaussian dist. and ideal reliab. r=0.98297, R^2=0.95044
Between Weibull dist. and ideal reliab. r=0.98999, R^2=0.98008
```

This shows that the data are better fitted with Weibull distribution than with Gaussian distribution, although the correlation coefficients seem to be similar, $\Delta r = 0.00702$. If we compare them with the coefficient of determination, $\Delta R = 0.02964$. Note that in the case of linearity, it is customary to use the correlation coefficient rather than the coefficient of determination to determine the goodness of fit (Figure 8.2).

Observant readers may notice that the correlation coefficient in the first part of this code is $r=0.986339$, but in the second part, the correlation coefficient becomes 0.98999. This is because the definition of the two is different. The former is the "post-log correlation

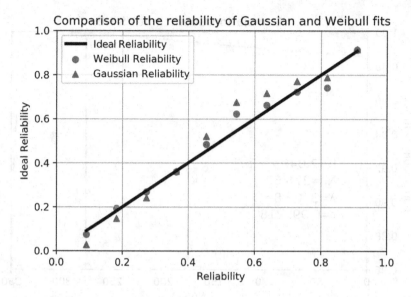

FIGURE 8.2 Comparison chart of fitting reliability between Gaussian and Weibull distribution.

coefficient", i.e., the correlation coefficient between $Y_i = \ln(\ln(1/p_i))$ and $X_i = \ln(x_i)$ as defined by (8.12), while the latter is the correlation coefficient between the "ideal" reliability before taking the logarithm (8.10) and the data considered as a Weibull distribution. Therefore, the latter correlation should be "more reliable" in this sense.

Example 8.2

Now, use the data of Gao (1986) to find the values of the three parameters of the Weibull distribution using the ZT Gao method. Simply modifying the input data immediately produces the following results:

```
N= [350, 380, 400, 430, 450, 470, 480, 500, 520, 540, 550, 570, 600, 610, 630, 650,
670, 730, 770, 840]
Nav=, 557.0,s= 132.152, Nm= 545.0
r= 0.999218,bm= 2.040,λ= 320.98,N0= 276.60
```

From the calculation results, it can be seen that using the ZT Gao method to calculate the three parameters of the Weibull distribution actually maximizes the correlation coefficient. This indicates an improvement in the ZT Gao method. Although it may not seem significantly different from the results on page 142 of Gao (1986) ($N_0 = 300$, $b = 1.74$, $r = 0.998$), the differences in shape parameter b and safe life N_0 appear to be noteworthy (Figure 8.3).

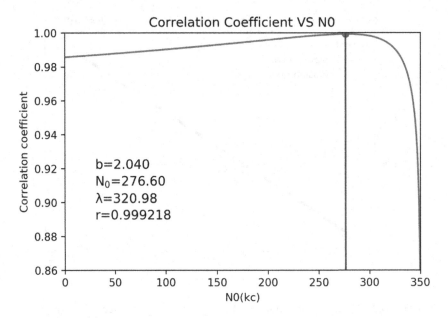

FIGURE 8.3   Another example by using the Zhentong Gao method.

Based on Mr. Gao's contribution to fatigue statistics, naming this new algorithm for estimating the three parameters of the Weibull distribution as the Zhentong Gao method is well-deserved. Although he himself does not necessarily agree, a related paper by the second author of this book has been publicly published, making this naming a proper noun (Xu, 2021). In fact, if Mr. Gao had not proposed to rewrite two of his original books (<Applied Statistics for Fatigue> (Gao,1986) and <Fatigue Reliability> (Gao & Xiong,2000)) into <Intelligent Fatigue Statistics>(Gao,& Xu, 2022), it is likely that this Zhentong Gao Method would have remained undiscovered until now.

8.2 APPLICATION OF ZHENTONG GAO METHOD: FITTING THE THREE-PARAMETER FATIGUE PERFORMANCE CURVE

8.2.1 Fitting the Three-Parameter Fatigue Performance Curve

In Section 7.4, three-parameter empirical formulas for fatigue performance were introduced. For example, for (7.70):

$$(S_{max}-S_0)^m N = C \tag{8.15}$$

Here, S_0, m, and c are material constants. The a-N curve of fatigue propagation is also a three-parameter equation:

$$N = C(a(N) - a_0)^m \tag{8.16}$$

where a_0 is the initial crack size, and C and M are material constants. For composite materials under constant loading, the E-N (elastic modulus-stress cycle) relationship can be expressed as follows:

$$N = C(E_0 - E(N))^m \tag{8.16a}$$

These three-parameter empirical formulas are commonly used for estimating fatigue life. Taking (8.15) as an example, taking logarithms on both sides:

$$lgN = lgC - mlg(S_{max} - S_0),$$

For convenience, let's define:

$$X = lgN, \ y = lg(S_{max} - S_0), \ a = lgC, \ b = -m \tag{8.17}$$

Then, we have:

$$X = a + by \tag{8.18}$$

Following the results in Section 7.4, we have:

$$b = L_{YX}/L_{YY} \tag{8.18a}$$

$$a = \tilde{X} - b\tilde{Y} \tag{8.18b}$$

where,

$$L_{XY} = \sum_{i=1}^{n} x_i y_i - (1/n)\left(\sum_{i=1}^{n} y_i\right) = L_{YX} \tag{8.19}$$

$$L_{YY} = \sum_{i=1}^{n} y_i^2 - (1-n)\left(\sum_{i=1}^{n} y_i\right)^2 = \sum_{i=1}^{n} y_i^2 - n\left(\tilde{Y}\right)^2 \quad (8.19a)$$

$$L_{XX} = \sum_{i=1}^{n} x_i^2 - (1/n)\left(\sum_{i=1}^{n} x_i\right)^2 = \sum_{i=1}^{n} x_i^2 - n\left(\tilde{X}\right)^2 \quad (8.19b)$$

$$r = [E(X-\mu_x)E(Y-\mu_y)]/\{E[(X-\mu_x)^2]\{E[(X-\mu_x)^2]\} = Cov(X,Y)/\sigma_x\sigma_y \quad (8.19c)$$

$$r = L_{XY}/(L_{YY}L_{XX})^{1/2} \quad (8.19d)$$

It can be seen that $\tilde{Y}$, L_{YY}, L_{XY} are all functions of S_0. Therefore, a, b, and r are also functions of S_0. To find S_0, we can maximize the correlation coefficient $|r|$, which means:

$$d|r|/dS_0 = 0, \text{ i.e.:} \qquad dr^2/dS_0 = 0.$$

Noticing that:

$$dr^2/dS_0 = 2r^2[(dL_{XY}/dS_0)/L_{XY} - (dL_{YY}/dS_0)/2L_{YY}]$$

therefore,

$$(dL_{XY}/dS_0)/L_{XY} - (dL_{YY}/dS_0)/2L_{YY} = 0 \quad (8.20)$$

To further simplify, we can introduce (Gao & Xiong, 2000):

$$L_{X0} = \ln10\left(dL_{XY}/dS_0\right)$$
$$= -\sum_{i=1}^{n} x_i/(S_i - S_0) - (1/n)\left(\sum_{i=1}^{n} x_i\right)\sum_{i=1}^{n}[1/(S_i - S_0)] \quad (8.21)$$

$$L_{Y0} = (\ln10/2)\left(dL_{YY}/dS_0\right)$$
$$= -\sum_{i=1}^{n} y_i/(S_i - S_0) - (1/n)\left(\sum_{i=1}^{n} y_i\right)\sum_{i=1}^{n}[1/(S_i - S_0)] \quad (8.21a)$$

From (8.20), we have:

$$L_{X0}/L_{XY} - L_{Y0}/L_{YY} = 0 \qquad (8.22)$$

It can be rewritten as:

$$H(S_0) = L_{X0}/L_{XY} - L_{Y0}/L_{YY} = 0 \qquad (8.23)$$

This equation needs to be solved using a computer, typically through the Newton-Raphson method. Although it may be a bit troublesome, it can be solved numerically. Once S_0 is obtained, we can calculate:

$$m = -b = -L_{YX}/L_{YY}, \quad C = 10^a = 10^{x~ - by~} = 10^{x~ + my~} \qquad (8.23a)$$

For equations like (8.16), similar methods can be applied to solve them. By defining various parameters as in (8.17), we can obtain:

$$Y = a + bx \qquad (8.24)$$

$$X = \lg N, \quad y = \lg(a(N) - a_0), \quad b = L_{yx}/L_{xx} = 1/m, \quad a = -\lg C/m \qquad (8.25)$$

To facilitate calculations in a computer, (8.21) and (8.19) can be rewritten as:

$$L_{X0} = \sum_{i=1}^{n}(x_i - \tilde{X})/(S_i - S_0); \quad L_{Y0} = \sum_{i=1}^{n}(y_i - \tilde{Y})/(S_i - S_0);$$

$$L_{XY} = \sum_{i=1}^{n} y_i(x_i - \tilde{X}) \qquad (8.26)$$

Example 8.3

Experimental results are shown in the table below (Gao & Xiong, 2000). lgN is the logarithm of the average fatigue life for each group. Calculate the S-N curve (Table 8.1).

TABLE 8.1   Experimental Date

Group No.	S_{max}/MPa	lgN
1	380	2.5933
2	353.6	2.8976
3	326.4	3.2201
4	299.2	3.8671

8.2.2 Using Excel to Solve Example 8.3

The core of solving this problem is to find the optimal S_0. How can this be achieved? Essentially, it still involves using the Newton-Raphson method to iteratively calculate S_0. It cannot be done automatically like Python, but human intervention can still be minimized. In Section 3.2.2, it was mentioned that Excel can be used to solve transcendental equations, but the specific steps were not explained. Here are the detailed steps for your reference. First, create Table 8.2 in Excel.

At this point, you only need to change the value of S_0 to get the desired value immediately, as shown in Table 8.3.

The "manual" calculations are relatively easy once Table 8.2 has been made, but it still takes some work to make the table. It is also necessary to draw a diagram based on the results (Figure 8.4 and Table 8.4).

8.2.3 Using Python and Zhentong Gao Method to Solve Example 8.3

It is quite clear that using Excel is simpler, but when it comes to solving and graphing complex problems like this one, it may become a bit challenging. Python programming can be a bit more complicated and may encounter some unexpected issues, but it provides better results, and graphing is not difficult. Most importantly, Python code can be that porting is relatively easy. Throughout this process, you will gain a deeper understanding of the similarities and differences between Var and L_{xx}. In Section 6.4, Python code for solving this problem was provided. With slight modifications and running it on Spyder, you can obtain the following results:

```
k(Number of dichotomy)= 11,E(precision)= 1e-07,S0=270.89
m= 2.143,C= 9.4570e+06,r= -0.99914
```

This result is also very close to the results of Excel and Gao's work (Gao, 2000), as shown in Table 8.5. The fitted curve can be expressed as $(S-270.89)^{2.1428} = 9.457 \times 10^6 (C)$ (Figure 8.5).

These three methods can be said to be very consistent; the only difference lies in the "calculation error" and the complexity of the calculation process, as well as whether there is generalizability or whether it can be replicated. There is no doubt that Python is superior from any perspective for such problems, so it will be used in the future.

As for whether we can use the ZT Gao method to solve such problems, the answer is yes. This is because, from a mathematical perspective, there is

TABLE 8.2 The Trial Calculation of S_0

Group	S_{max}/MPa	lgN	$S_i - S_0$	L_{X0}	L_{XY}	L_{Y0}	$lg(S_i - S_0)$	$lg^2(S_i - S_0)$	lg^2N
1	380	2.5933	109.11	-0.00505	-1.1235	0.002290635	2.03786	4.152892	6.7252
2	353.6	2.8979	82.71	-0.00298	-0.4731	0.001567218	1.91756	3.677029	8.39782
3	326.4	3.2201	55.51	0.00136	0.1317	-0.00078476	1.74437	3.042831	10.369
4	299.2	3.8671	28.31	0.025521	1.04903	-0.01186837	1.45194	2.108129	14.9545
Σ=		12.5784	275.64	0.018846	-0.4158	-0.00879528	7.15173	12.98088	40.4465
		LXX=	0.8925	LX0/LXY=	m	LY0/LYY=	-0.0453	$L_{YY}=$	0.19406
		S_0	H(S_0)	r	2.14271	C			
		270.89	1.259E-07	-0.99914		9.454E+06			

TABLE 8.3   Trial Result

S_0	$H(S_0)$	r	m	C
280	4.618E-04	−0.9984	1.74388	1.308E+06
260	−1.642E-04	−0.99867	2.58875	9.472E+07
270	−2.271E-05	−0.99913	2.18014	1.143E+07
275	1.447E-04	−0.99902	1.96678	3.909E+06
272	3.2284E-05	−0.99913	2.09572	7.456E+06
271	3.1229E-06	−0.99914	2.13807	9.235E+06
270.9	3.9669E-07	−0.99914	2.14229	9.434E+06
270.89	1.2588E-07	−0.99914	2.14271	9.454E+06

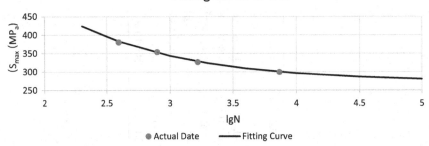

FIGURE 8.4   Fitting curve of S-N.

TABLE 8.4   Actual Value and Fitting Value of S-N Curve

Actual lgN	2.5933	2.8979	3.2201	3.8671		S_0	m	lgC
Smax/MPa	380	353.6	326.4	299.2		271.89	2.143	6.9756
Fitting lgN	2.3	2.6	3	3.3	3.6	4	4.5	5
S_{max}/MPa	423.98	382.07	343.57	323.819	309.51	296.36	286.19	280.25

TABLE 8.5   Results of Three Different Calculation Methods

	S_0	m	C	r
in (Gao & Xiong, 2000) P318	270.89	2.1425	9.445E+06	−0.9991
using Excel	270.89	2.14271	9.454E+06	−0.9991
using Python	270.89	2.1428	9.457E+06	−0.9991

no significant difference between the three-parameter Weibull distribution and the three-parameter fatigue performance curve problem. You only need to use (8.17) and (8.18), without the need for further analysis of the expression for the correlation coefficient r. In the equation X=a+bY, where X=lgN, Y=lg(S_{max}−S_0), a=lgC, and b=−m, S_0 is unknown. However, finding S_0 is

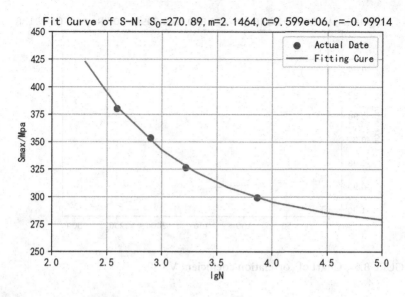

FIGURE 8.5   S-N fitting curve obtained using Python.

similar to finding N_0 in Section 8.1; it can be determined by maximizing the correlation coefficient (Xu, 2021). By modifying the ZT Gao method code introduced in Section 8.1.3, the following results can be obtained:

```
r= -0.999138,  m= 2.146,S0= 270.89,C= 9.599e+06
```

Now, let's analyze the results obtained by this method and compare them with the results obtained by the previous three methods. It is not difficult to see that S0 is almost "indistinguishable" from the others, with the only significant difference being in the value of C, which has a relatively larger difference (but the relative error is still within 1.7%). The relative errors of the other parameters are all less than one-thousandth. The relatively larger error in C is due to its exponential nature, which can amplify errors (Figures 8.6 and 8.7).

Example 8.4

Given the a-N data of a certain experiment, the a-N curve equation is to be determined (Gao & Xiong, 2000) (Table 8.6).

Solution

This problem can be solved first using the analytical method. It must be noted that (8.18) and (8.24) are formally different with respect to

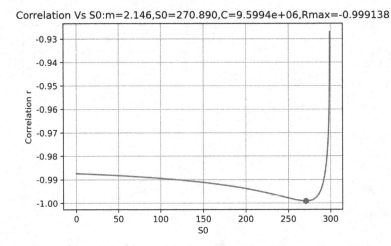

FIGURE 8.6   Chart of correlation coefficient VS S0.

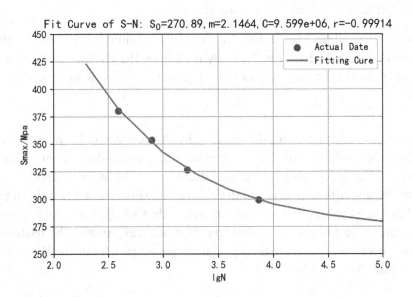

FIGURE 8.7   S-N fitting curve obtained using the ZT Gao method.

TABLE 8.6   An Experimental Date

Group	1	2	3	4	5	6	7
a/mm	0.4978	0.642	0.6655	0.7976	0.9322	1.0998	1.2929
N/h	4,800	5,200	5,600	6,000	6,400	6,800	7,200

calculating a, b and m, C because the equations change form. In other words, note that at this point, $x = \lg N$, $Y = \lg(a(N) - a_0)$, (8.25) became:

$$m = L_{XX} / L_{YX} = 1/b, C = 10^{-ma} \text{ and } a = \tilde{x} - b\tilde{y} \qquad (8.27)$$

As you can see, you only need to replace S with a, and the input data will naturally be different. Now, the transcendental equation (8.23) is no longer solving for S_0 but for a_0, but the solution method remains the same. The parameters to be changed in the end are m and C, but r remains "the same". Therefore, you must appropriately modify and improve the code to ensure that the computed results are error-free! Readers can also give it a try; this is the real "challenge". Understanding the principles and applying them to solve practical problems is essential. The results are as follows:

```
k(number of dichotomy)= 14,E(precision)= 1e-06,a0=0.17340
m= 0.34286,C= 6.990e+03,r= 0.99203
```

As a result, the best-fitted curve is: $N = 6990(a-0.1734)^{0.3429}$ (Figure 8.8).

The example can also be done by the ZT Gao method, except that the equation obtained after taking the logarithm is somewhat

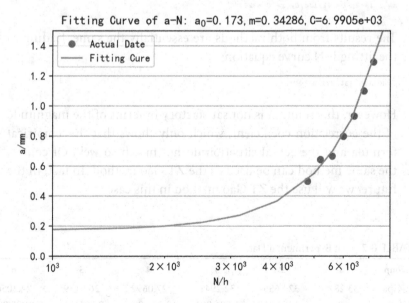

FIGURE 8.8   a-N fitting curve obtained using the analytical method.

different in form and also requires appropriate modifications to the code to run the result as:

```
r= 0.99203, a0= 0.169, m= 0.345, C= 6.9814e+03
```

The results obtained are almost no different from the analytical method, and the fitted graphs do not change much, so they are not given here.

Example 8.5

Given the test results for a composite material, determine the E-N fitting curve (Gao & Xiong, 2000) (Table 8.7).

From a mathematical perspective, (8.16a) and (8.16) are identical in form, meaning it is "the same" as Example 2. However, it is essential to note that in this case, $Y=\lg(E_0-E)$ rather than $E-E_0$. Simply modify the input raw data, but pay attention to some details, especially noting that the first data point approximately implies $E_0=33.2871$. With this data, the remaining values for m, C, and r can be easily calculated. It can also be done using Excel (Figure 8.9 and Table 8.8).

Now, let's look at the results obtained using the analytical method. When executed in Spyder, the results are as follows:

```
m= 1.1344, C= 423134.0, r= 0.88241
```

The results from both methods are essentially the same, leading to the fitting E-N curve equation:

$$N=423134(33.2871-E)^{1.1344}.$$

However, this solution is not satisfactory in terms of the magnitude of the correlation coefficient, which only shows that this empirical formula and the actual situation do not match so well. Of course, the same method can be used as the ZT Gao method. In fact, in the future, we will use the ZT Gao method in this case.

TABLE 8.7 An Experimental Date

Group	1	2	3	4	5	6
E/Gpa	33.2871	32.8655	32.2941	27.0692	26.5139	24.3056
N/c	0	126,001	1,526,001	2,275,000	2,391,500	4,055,000

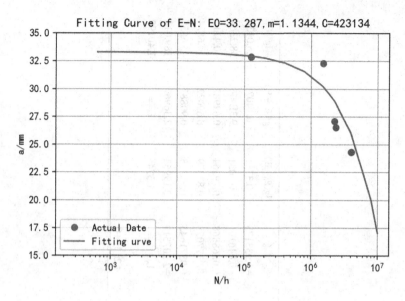

FIGURE 8.9　E-N fitting curve obtained using the analytical method.

The results of using the ZT Gao method are as follows:

r= 0.88215,　E0= 33.29,　m= 1.137,　C= 4.2097e+05

As in Example 8.2-2, the results obtained are almost no different from the analytical method, and the fitted graphs do not change much, so they are not given here.

8.3 MAXIMUM LIKELIHOOD ESTIMATION AND GENERALIZED ZHENTONG GAO METHOD

As mentioned in Section 7.5, the maximum likelihood estimation(MLE) for the three parameters of the Weibull distribution is challenging, with complex derivations and calculations. However, with the assistance of computers, it can be solved. Considering that the essence of the ZT Gao method is to directly find the extremum of a function using a "brute force" approach, and the MLE involves finding the extremum of the likelihood function, we can use the idea of ZT Gao method to find the extremum of the Weibull distribution likelihood function [Xu, 2023]. According to the definition of the likelihood function in Section 7.5, we can easily obtain the logarithm of the likelihood function for the Weibull distribution:

$$LL = InL = n*\ln(b/\lambda) + (b-1)\sum_{i=1}^{n}\ln[(x_i - x_0)/\lambda] - \sum_{i=1}^{n}[(x_i - x_0)/\lambda]^b$$

$$(8.28)$$

TABLE 8.8 Using Excel to Calculate the Parameters of E-N Curve

Group	E/GPa	N/c	$\lg N$	E_0-E_i	L_{X0}	L_{XY}	L_{Y0}	$\lg(E_0-E_i)$	$\lg^2(E_0-E_i)$	$\lg^2 N$
1	32.8655	126001	5.100374	0.4216	−2.4315	0.38453	−1.933175	−0.3751	0.1407	26.0138
2	32.2941	1526001	6.183555	0.993	0.05845	−0.0002	−0.446101	−0.0031	9.3E-06	38.2364
3	27.0692	2275000	6.356981	6.2179	0.03723	0.1837	0.0568868	0.79364	0.62987	40.4112
4	26.5139	2391500	6.37867	6.7732	0.03738	0.21032	0.0577078	0.83079	0.69022	40.6874
5	24.3056	4055000	6.607991	8.9815	0.05372	0.45997	0.0571643	0.95335	0.90887	43.6655
Σ=			30.62757	23.3872	−2.2448	1.23834	−2.207517	2.19964	2.36967	189.014
			$L_{XX}=$	1.40473	L_{X0}/L_{XY}	−1.8127	$L_{X0}/L_{YY}=$	−1.5746	$L_{YY}=$	1.40199
			E_0	$H(S_0)$	r	m	C			
			33.2871	2.382E-01	0.88241	1.13436	4.213E+0			

Now, instead of separately taking partial derivatives of the three parameters b, λ, and x_0 in the equation above and setting them to zero, we can directly use a brute force approach to find the three parameters that maximize (8.28). This method computes what can be called the "generalized Zhentong Gao method (can be abbreviated as G_ZT Gao method)." However, it's important to note that the prerequisite for using this method is to have a rough estimate of the ranges of variation for these three parameters, which is perhaps the challenging aspect of using the generalized Zhentong Gao method. Fortunately, because the range of variation for the parameter x_0 is already determined and the ranges for the shape parameter b and scale parameter λ can be relatively easily determined based on the results of the Zhentong Gao method, this method is referred to as the "generalized Zhentong Gao method". By directly referencing the results of Example 3 in reference (Xu, 2023), the following results are obtained:

```
X= [3956.42, 4004.18, 4091.61, 4355.05, 4355.4, 4376.01, 4391.79, 4487.68, 4487.68,
    4736.67, 4736.67, 4939.85, 4963.62, 5220.19, 5353.41, 5372.72, 5418.04, 5444.11,
    5603.17, 5698.1, 5746.17, 5843.52, 6175.14, 6197.41, 6249.69, 6279.76, 6279.76,
    6572.74, 6740.48, 6887.65, 7183.09, 7209, 7209, 7209, 7209, 7366.4, 7581.64,
    7581.64, 7581.64, 7645.59, 8246, 8599.7, 8713.97, 8936.34, 9044.22, 9197.45,
    9511.73, 9754.47, 9967.45, 10136.31, 10172.88, 10172.88, 10308.04, 10395,
    10609.23, 10609.23, 10788.97, 10879.97, 10971.75, 11594.41, 11990.59, 12237.31,
    12400.31, 12400.31, 12550.01, 13198.73, 13947.78, 15557.12, 17646.12, 19848.23,
    23199.07]
```

Through the ZT Gao method, corresponding estimates for the three parameters are obtained. With these estimates, the ranges of variation and step sizes for these three parameters are determined. Then, the generalized ZT Gao method is used to obtain the corresponding estimates. Finally, a comparison is made with the MLE estimates, resulting in Table 8.9, which compares the results obtained by various estimation methods:

TABLE 8.9   Comparison of Results Obtained by Various Estimation Methods

	b	λ	N_0	R^2	LL
ZT Gao	1.146	4,957.7	3,813.0	**0.99391**	−667.738
G_ZT Gao	1.190	4,800.0.	3,877.3	0.99297	**−667.054**
MLE	1.139	4,946.1	3,821.5	0.99378	−667.707

Data in bold indicates optimal value.

The estimated values of three parameters of MLE in the table are from Yan et al. (2005). From the data in Table 8.9, it can be found that the estimated values of the three methods, whether LL values or three parameters, are quite close, so it is difficult to judge which method has a better estimated value, and other fitting standards may be needed. In any case, it can still be said that the results obtained by the ZT Gao method are satisfactory. At the same time, the results obtained by the G_ZT Gao method are almost the same as those obtained by MLE, and its corresponding LL value is still the largest. Once again, it is proven that the G_ZT Gao method is effective.

Finally, histograms of the data and fitted PDFs of different Weibull distributions can be obtained, as shown in Figure 8.10.

From Figure 8.10, it can be seen that the PDF figures of Weibull distribution estimated by three different methods are almost indistinguishable. This also shows that in a sense, the same Weibull distribution can correspond to almost infinite three different parameter combinations within a certain error range. Of course, in any case, it is impossible to combine its location parameter greater than the minimum value of the given data.

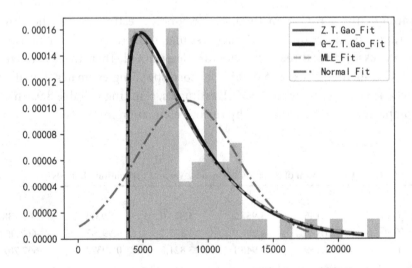

FIGURE 8.10 Fitting diagram of data histogram and different WD's PDF.

REFERENCES

Gao ZT (1986), *Fatigue Applied Statistics*, National Defense Industry Press, Beijing.

Gao ZT & Xiong JJ (2000), *Fatigue Reliability*, Beihang University Press, Beijing, p. 155, 318, 319, 320.

Gao ZT & Xu JJ (2022), *Intelligent Fatigue Statistics*, Beihang University Press, Beijing.

Trivedi K S (2015), *Probability and Statistics with Reliability, Queuing, and Computer Science Applications*, Electronic Industry Press, Beijing, p. 597, 602.

Xu JJ (2021), Zhentong Gao Method in Intelligentization of Statistics in Fatigue, *Journal of Beijing University of Aeronautics and Astronautics*, 47(10): 2024–2033. doi: 10.13700/j.bh.1001-5965.2020.0373

Xu JJ (2023), Generalized Zhentong Gao Method for Estimating Three Parameters of Weibull Distribution, *SCIREA Journal of Computer*, 8(2). doi: 10.54647/computer520345

Yan XD, Ma X, Zheng R, & Wu L (2005), Comparison of the Parameters Estimation Methods for three Parameter Weibull Distribution, *Journal of Ningbo University*, 18(3): 301–305.

CHAPTER 9

Several Commonly Used Testing Methods

9.1 CHI-SQUARE (χ^2) TEST

9.1.1 Probability Density Function of Chi-Square Distribution

The χ^2 distribution is of significant importance in mathematical statistics. It was first introduced by Abbe in 1863 and later derived independently by Helmert and one of the founders of modern statistics, C. K. Pearson, in 1875 and 1900, respectively.[1] It arises from v mutually independent Gaussian standard random variables U_1, U_2,..., U_v. A new random variable χ^2 is defined as the sum of squares of these variables:

$$\chi^2 = U_1^2 + U_2^2 + \ldots + U_n^2 = \sum_{i=1}^{v} U_i^2 \tag{9.1}$$

This random variable follows the χ^2 distribution.

Now, let's find the PDF of the χ^2 distribution. To do this, we'll first find its moment generating function (MGF). Since the random variables U_1, U_2, ..., U_V are mutually independent:

$$M_{\chi^2}(\theta) = M_{U_1^2}(\theta)\, M_{U_2^2}(\theta) \ldots M_{U_v^2}(\theta) \tag{9.2}$$

It's noted that (Gao, 1986):

$$M_{U_1^2}(\theta) = M_{U_2^2}(\theta) = \ldots = M_{U_v^2}(\theta) = (1 - 2\theta)^{-1/2} \tag{9.3}$$

DOI: 10.1201/9781003488477-12

Therefore:

$$M_{\chi^2}(\theta) = (1 - 2\theta)^{-v/2} \tag{9.4}$$

In Section 3.3, the PDF of gamma distribution is given as: $f(x) = [\beta^\alpha/\gamma(\alpha)]x^{\alpha-1}\exp(-\beta x)$, and its moment generating function is:

$$M_\xi(\theta) = (1 - \theta/\beta)^{-\alpha} \tag{9.5}$$

Comparing (9.4) and (9.5), we can derive:

$$\alpha = v/2 \text{ and } \beta = 1/2 \tag{9.6}$$

According to the uniqueness theorem of moment generating functions, their PDFs should also be the same. Therefore, the PDF of the χ^2 distribution is:

$$f_v(x) = [(1/2)^{v/2}/\Gamma(v/2)]x^{v/2-1}\exp(-x/2) \tag{9.7}$$

Similarly, from Section 3.3, it is known that the mathematical expectation and variance of the gamma distribution are:

$$E(\chi^2) = \alpha/\beta = v; \text{ Var}(\chi^2) = \alpha/\beta^2 = 2v \tag{9.8}$$

This indicates that v is a parameter of the χ^2 distribution, and in this context, it is referred to as the "degrees of freedom". It emphasizes the number of mutually independent variables. As we will see later, it has a "physical meaning", where an increase in "constraints" leads to a decrease in degrees of freedom. Thus, the following theorem can be obtained (Gao, 1986):

Theorem 9.1

If U_1, U_2,..., U_v are mutually independent standard normally distributed random variable, then $\sum_{i=1}^{v} U_i^2$ follows the χ^2 distribution (9.7), and v is its degree of freedom.

Similar to the above method for the moment mother function, it is easy to obtain.

Theorem 9.2

If χ_1^2 and χ_2^2 are mutually independent χ^2 variables, and their degrees of freedom are v_1 and v_2, respectively, then $\chi_1^2 + \chi_2^2$ is also a random variable distributed according to χ^2 with degrees of freedom $v_1 + v_2$.

This theorem is also not difficult to generalize to the case of a finite number of mutually independent χ^2 variable. The following Theorem 9.3 is obtained from the "uniqueness theorem". This is to prepare for proving Theorem 9.4.

Theorem 9.3

If Ω_1, Ω_2 are two mutually independent random variables, Ω_1, + Ω_2 is a χ^2 variable (the word "random" is omitted here, and if not specified, the word "variable" will be referred to as "random variable" in the future). The degree of freedom is $v_1 + v_2$. And if Ω_1 is a χ^2 variable with degree of freedom v_1, then Ω_2 must be a χ^2 variable with degree of freedom v_2.

Theorem 9.4

If s_x^2 is a sample variance of size n drawn from the population of $N(\mu; \sigma^2)$, then $s_x^2(n-1)/\sigma^2$ is a χ^2 variable with degree of freedom $v = n - 1$.

Proof: Noting a χ^2 variable with degree of freedom n, χ^2 follows from equation (9.1) that:

$$\chi^2 = \sum_{i=1}^{n} U_i^2 = \sum_{i=1}^{n} \left[(X_i - \mu)/\sigma\right]^2 = \sum_{i=1}^{n} \left[(X_i - \tilde{X})/\sigma + (\tilde{X} - \mu)/\sigma\right]^2$$

$$= s_x^2(n-1)/\sigma^2 + \left[(\tilde{X} - \mu)/(\sigma/n^{1/2})\right]^2$$

It should be noted here that:

$$\sum_{i=1}^{n} \left[2(X_i - \tilde{X})(\tilde{X} - \mu)/\sigma^2\right] = 0$$

Because $\left[(\tilde{X} - \mu)/(\sigma/n^{1/2})\right]^2$ is a χ^2 variable with degree of freedom 1, according to Theorem 9.3, it is immediately clear that $s_x^2(n-1)/\sigma^2$ is a χ^2 variable with degree of freedom $v = n - 1$, thus, q.e.d.

When the degree of freedom v is determined, the PDF (9.7) of the χ^2 variable is determined, and its probability distribution is also determined, and the normal distribution can be calculated as one needs the probability. If for χ^2 greater than a certain specified χ_a^2 probability is equal to (Figure 9.1):

$$P\left(\chi^2 > \chi_a^2\right) = \int_{\chi_a^2}^{\infty} \left[(1/2)^{v/2}/\Gamma(v/2)\right] x^{\frac{v}{2}-1} \exp(-x/2)dx \qquad (9.9)$$

9.1.2 Principle of Chi-Square Test

It has already been pointed out that the population standard deviation σ characterizes whether the quality of the product is uniform or not. Therefore, it is necessary to do a test on the mean squared deviation of the sample and whether it conforms to the population standard deviation, and chi-square test is needed. This method does not have special requirements for the sample size (Figure 9.2).

Let s_x^2 be the standard variance of a sample of capacity n drawn from a population of $N(\mu,\sigma^2)$: $s_x^2 = \sum_{i=1}^{v}\left[\left(X_i - \tilde{X}\right)^2/(n-1)\right]$

As we saw in the previous section, a χ^2 variable with $n-1$ degrees of freedom can be defined as:

$$\chi^2 = s_x^2(n-1)/\sigma^2 \qquad (9.10)$$

Schematic diagram of PDF of χ^2 Distribution

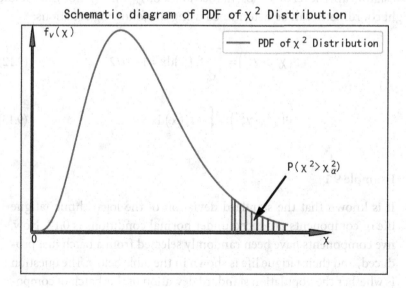

FIGURE 9.1   Schematic diagram of χ^2-distribution's PDF.

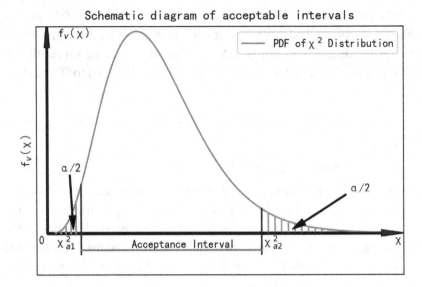

FIGURE 9.2 Schematic diagram of χ^2-distribution's PDF and significance.

In practical situations, it is necessary to know the population variance, denoted as $\sigma = \sigma_0$. In this case, the equation can be modified as follows:

$$\chi^2 = s_x^2 (n-1)/\sigma_0^2 \tag{9.11}$$

If a significance level α is chosen, the values of $\chi_{\alpha_1}^2$ and $\chi_{\alpha_2}^2$ for the left and right coordinates can be determined using the following equations:

$$P(\chi^2 > \chi_{a_1}^2) = \int_{\chi_{a_1}^2}^{\infty} f_v(x)dx = 1 - \alpha/2 \tag{9.12}$$

$$P(\chi^2 > \chi_{a_2}^2) = \int_{\chi_{a_2}^2}^{\infty} f_v(x)dx = \alpha/2 \tag{9.13}$$

Example 9.1

It is known that the standard deviation of the logarithmic fatigue life of components produced under normal conditions is 0.12. Now, five components have been randomly selected from a batch just produced, and their fatigue life is shown in the table below. The question is whether the population standard deviation of this batch of components meets the requirement (Gao, 1986).

The results listed in Table 9.1 can be made by Excel, just need to input the fatigue life. The rest is done by Excel's own functions.:

Here is a brief explanation of how this table is obtained. In Excel, in the first line in the first five fatigue life data, the second line can be used in the Excel formula LOG10 () automatically and do not have to calculate one by one. Therefore, you only need to calculate one and then use the COPY function to calculate all the rest. Then, use STDEV.S() to get the sample $\left(x_i^2\right)$ standard deviation s. And according to equation (9.10), we can get χ^2. Finally, use CHIINV(0.95,4) to get the value of χ_{a1}^2; the first parameter is determined by the significance α, that is, by equation (9.12) →1 − $\alpha/2$ = 0.95; and the second parameter is the degree of freedom v = n − 1 = 5 − 1 = 4. Again, the value of χ_{a2}^2 is obtained by CHIINV(0.05,4), when $\alpha/2$ = 0.05, while the degrees of freedom are the same.

Finally, the table shows that: $\chi_{a1}^2 < \chi^2 < \chi_{a2}^2$ and the value of χ^2 lies in the acceptable interval $(\chi_{a1}^2, \chi_{a2}^2)$; so this batch of samples is in compliance with the requirements.

9.2 A FEW EXAMPLES OF APPLICATION OF CHI-SQUARE TEST TO FATIGUE STATISTICS

9.2.1 Interval Estimation of Normal Population Standard Deviation

Significance level pertains to a "point", meaning when the population standard deviation is already known whether the sample standard deviation is "consistent" with this known point. Confidence level, on the other hand, is an estimate of the "unknown" population standard deviation. Mathematically or formally, they are the same, but they have different "physical meanings". (Figure 9.3)

Hence:

$$P\left(\chi^2 > \chi_{\gamma 1}^2\right) = \int_{\chi_{\gamma 1}^2}^{\infty} f_v(x)dx = 1-(1-\gamma)/2 = (1+\gamma)/2 \qquad (9.14)$$

TABLE 9.1   Data of a Fatigue Life

Fatigue Life: N$_i$(kc)	108	120	160	170	258	s	s²
x$_i$ = lgN$_i$	2.033424	2.079181	2.20412	2.230449	2.41162	0.148098	0.021933
x$_i^2$	4.134812	4.322995	4.858145	4.974902	5.81591	σ²	χ²
Significance: α = 10%	$\chi_{\alpha1}^2$ =	0.710723	$\chi_{\alpha2}^2$ =	9.487729		0.0144	6.092516

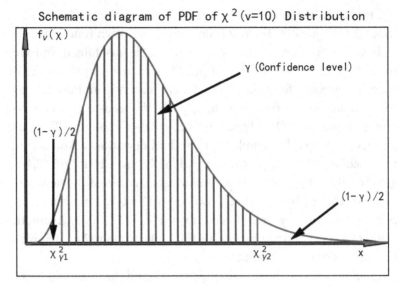

FIGURE 9.3   Schematic diagram of χ^2-distribution ($v = 10$)'s PDF and confidence.

$$P\left(\chi^2 > \chi^2_{\gamma_2}\right) = \int_{\chi^2_{\gamma_2}}^{\infty} f_v(x)dx = 1 + (1-\gamma)/2 = (1-\gamma)/2 \qquad (9.15)$$

It can be seen that when $1 - \gamma = \alpha$, these two equations are the same as (9.12) and (9.13):

$$\chi^2_{\gamma_1} = \chi^2_{\alpha_1}; \quad \chi^2_{\gamma_2} = \chi^2_{\alpha_2} \qquad (9.16)$$

This shows that $\gamma = 1 - \alpha$ is equivalent in mathematics, and the so-called look-up table is similar, just the opposite, but the "physical meaning" is different. That is, confidence interval of χ^2 is

$$\left(\chi^2_{\gamma_1}, \chi^2_{\gamma_2}\right), \quad \chi^2_{\gamma_1} < \chi^2 < \chi^2_2; \chi^2_{\gamma_1} < s^2(n-1)/\sigma^2 < \chi^2_{\gamma_2}$$

and,

$$s\left[(n-1)/\chi^2_{\gamma_2}\right]^{1/2} < \sigma < s\left[(n-1)/\chi^2_{\gamma_1}\right]^{1/2} \qquad (9.17)$$

This is the confidence interval for the population standard deviation.

Example 9.2

At a certain stress level, the standard deviation of fatigue life for ten specimens was measured to be 0.143. Determine a confidence interval for the population standard deviation at a 95% confidence level.

Solution

When $\gamma = 95\%$, we have:

$$P\left(\chi^2 > \chi^2_{\gamma_1}\right) = (1+\gamma)/2 = 97.5\%, \ P\left(\chi^2 > \chi^2_{\gamma_2}\right) = (1-\gamma)/2 = 2.5\%$$

Similarly, using Excel's built-in functions to calculate these values (without the need for table lookup), and replacing all manual calculations, the following results can be obtained:

As a result, the confidence interval for the population standard deviation of fatigue life at a confidence level of 95% is (0.098, 0.261) (Table 9.2).

9.2.2 Unbiased Estimator of Normal Population Standard Deviation

In Section 4.3.1, we discussed that an unbiased estimator of the population standard deviation is given by (Gao, 1986):

$$\hat{\sigma}^2 = s^2 = \left[1/(n-1)\right] \sum_i \left(x_i^2 - \tilde{X}\right)^2 = \left[1/(n-1)\right] \left[\sum_i x_i^2 - n\left(\sum_i x_i\right)^2\right]$$

Strictly speaking, s is not a truly unbiased estimator because it does not satisfy the unbiased requirement $E(s_z) = \sigma$. Notice that:

$$\chi^2 = s_x^2(n-1)/\sigma^2 \text{ and } s_x = \left[\sigma/(n-1)^{1/2}\right]\chi \tag{9.18}$$

therefore:

$$E(s_x) = E\{[\sigma/(n-1)^{1/2}]\chi\} = [\sigma/(n-1)^{1/2}]E(\chi) \tag{9.19}$$

TABLE 9.2   Result of Example 9.2

$\text{CHINV}(0.975,9) = \chi^2_{\alpha_1} =$	2.700389
$\text{CHINV}(0.025,9) = \chi^2_{\alpha_2} =$	19.02277
$\left[(s*(n-1)/\chi^2_{a_1}\right]^{1/2} =$	0.261062
$\left[(s*(n-1)/\chi^2_{a_2}\right]^{1/2} =$	0.09836

Using (9.7) and noting the properties of the gamma distribution (refer to Section 3.3.2):

$$E(\chi) = \int_0^\infty x^{1/2} f_v(x) dx = \int_0^\infty x^{1/2} \left[(1/2)^{v/2} / \Gamma(v/2) \right] x^{v/2-1} \exp(-x/2) dx$$

$$= \int_0^\infty \left[\left(\frac{1}{2} \right)^{v/2} / \Gamma(v/2) \right] x^{(v+1)/2-1} \exp(-x/2) dx$$

$$= 2^{1/2} \Gamma[(v+1)/2]/\Gamma(v/2) \int_0^\infty \left[(1/2)^{(v+1)/2} / \Gamma[(v+1)/2] \right] x^{(v+1)/2-1} \exp(-x/2) dx$$

$$= 2^{1/2} \Gamma[(v+1)/2]/\Gamma(v/2)$$

$$(9.20)$$

and,

$$E(s_x) = [\sigma/(n-1)^{1/2}]\, 2^{1/2} \Gamma[(v+1)/2]/\Gamma(v/2) = [2/(n-1)]^{1/2} \Gamma(n/2)/\Gamma[(n-1)/2]\sigma$$

$$(9.21)$$

Therefore, the estimated amount of population standard deviation is:

$$\hat{\sigma} = [(n-1)/2]^{1/2} \Gamma[(n-1)/2]/\Gamma(n/2) s \qquad (9.22)$$

It can also be written as:

$$\hat{\sigma} = ks \qquad (9.23)$$

where,

$$k = [(n-1)/2]^{1/2} \Gamma[(n-1)/2]/\Gamma(n/2) \qquad (9.24)$$

k is called "standard deviation correction coefficient".

Similarly, using the GAMMA function that comes with Excel, the correction coefficient of any n can be easily given as shown in Table 9.3:

It is not difficult to see that when $n \to \infty$, $k \to 1$. This is also natural because the χ^2 distribution of n tends to infinity, and it becomes a normal distribution, so there is no need to correct it. Note that for any distribution $\tilde{\mu} = \tilde{X}$, while for normal distribution: $\tilde{\sigma} = ks$. Therefore, "standard fraction" is available:

$$x_p = \hat{\mu} + u_p \hat{\sigma} = \tilde{X} + k u_p s \qquad (9.25)$$

TABLE 9.3   Standard Deviation Correction Coefficient

n	5	6	8	10	15	20	30	50
k	1.063846	1.050936	1.036237	1.028109	1.018002	1.013239	1.008656	1.005115

Example 9.3

Calculate fatigue life with 99.9% reliability by using the data in Table 7.1.

It can be seen from Table 9.2 that $\tilde{X} = 2.167$; $s = 0.05$; as for the reliability of 99.9%, according to Excel we have that:

NORM.S.INV(0.001)=	−3.090232

Therefore:

$$u_p = -3.09, \text{ i.e., } x_p = 2.167 + 1.028*(-3.09)*0.05 = 2.008$$

and,

$$\lg N_p = x_p, \text{ i.e., } N_p = 10^{2.008} = 102(kc)$$

9.2.3 Chi-Square Test Method for Statistical Hypothesis of Population Distribution

Determining whether a population distribution belongs to the Gaussian distribution or Weibull distribution is a very important problem. The previous "graphical method" is relatively rough. Now, we will introduce the χ^2 test method, which is more precise (Figure 9.4).

To perform this test, a large sample is needed, n > 50, preferably 100. The sample mean $\tilde{X}$ and standard deviation s can be used to replace the population mean μ and standard deviation σ. At this point, n observations can be divided into m groups, represented by $v_1, v_2, ..., v_M$, which represent the actual observed frequencies of each group. $E_1, E_2,..., E_m$ represent the theoretical frequencies of each group. Hence:

$$\eta = \sum_{i=1}^{m} \left[(E_i - v_i)^2 / E_i \right]$$

$$= (E_1 - v_1)^2 / E_1 + (E_2 - v_2)^2 / E_2 + ... + (E_m - v_m)^2 / E_m$$

(9.26)

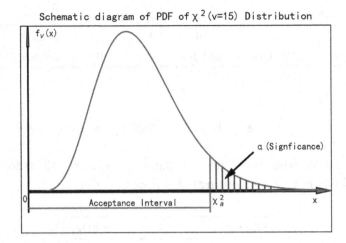

FIGURE 9.4 Schematic diagram of χ^2-distribution (v = 10)'s PDF and unilateral significance.

It is obvious that η is also a random variable, similar to (9.1), and it can be proven that when n is large, it is close to the χ^2 distribution. In addition, each group should have an actual frequency of more than 5 when grouping. Also, note that if the PDF has j undetermined parameters (such as μ and σ), then the corresponding χ^2 curve has degrees of freedom:

$$v = m - 1 - j \tag{9.27}$$

In this way, η can be regarded as χ^2 variable, and the value of η can be obtained by sampling once. Then, when the corresponding χ^2 value is obtained by Excel for a given significance α and degree of freedom v, it can be known whether it is in the acceptance range or not. It should be noted here that, unlike the previous significance test, the lower limit is zero, as shown in the following:

Example 9.4

Fatigue life N(105kc) of 100 specimens (Gao,1986) measured at the same stress level as shown in Table 9.4:

Divide the 100 specimens into the following nine groups, as shown in the following figure. It should be noted that such grouping is somewhat arbitrary. You can try different grouping methods:

It can be seen that the mean and standard deviation calculated by these two tables are somewhat different. The main reason is that

TABLE 9.4 Fatigue Life N(10^3 kc)

3.08	3.26	3.32	3.48	3.49	3.56	3.69	3.7	3.78	3.79
3.8	3.87	3.95	4.07	4.08	4.1	4.12	4.2	4.24	4.25
4.28	4.31	4.31	4.36	4.54	4.58	4.6	4.62	4.63	4.65
4.67	4.67	4.72	4.73	4.75	4.77	4.8	4.82	4.84	4.9
4.92	4.93	4.95	4.96	4.98	4.99	5.02	5.03	5.06	5.08
5.06	5.1	5.12	5.15	5.18	5.2	5.22	5.38	5.41	5.46
5.47	5.53	5.56	5.6	5.61	5.63	5.64	5.65	5.68	5.69
5.73	5.82	5.86	5.91	5.94	5.95	5.99	6.04	6.08	6.13
6.16	6.19	6.21	6.26	6.32	6.33	6.36	6.41	6.46	6.81
7	7.35	7.82	7.88	7.96	8.31	8.45	8.47	8.79	9.87
N~=	5.315	STDEV.S=	1.28919						

Table 9.5 is calculated by "grouping". There is a simplification process here, that is, the median of the group is used instead of the mean of the group, so there is some error.

Here, we will explain the method of calculating the theoretical frequencies. First, we need to calculate the "standard scores". For example: $u_1 = (N_1 - \mu)/\sigma = (3.0 - 5.36)/1.31 = -1.8$,

$$u_2 = (N_2 - \mu)/\sigma = (3.8 - 5.36)/1.31 = -1.19$$

It can also be found by Excel. $P(N > N_1) = 1 - \text{NORM.S.DIST}(-1.8, \text{TRUE}) = 0.9641$ and $P(N > N_2) = 1 - \text{NORM.S.DIST}(-1.19, \text{TRUE}) = 0.8828$

Then, the probability that falls within the interval (N1, N2) is calculated as:

TABLE 9.5 Fatigue Life is Grouped to Obtain Various Data.

Group Number : i	1	2	3	4	5	6	7	8	9
Lower interval limit of group	3	3.8	4.6	5.4	6.2	7	7.8	8.6	9.4
Median of group $N_i(10^5c)$	3.4	4.2	5	5.8	6.6	7.4	8.2	9	9.8
Actual frequency $v_i(c)$	10	16	32	24	8	2	6	1	1
Grouping Life v_{iNi} (10^5c)	34	67.2	160	139.2	52.8	14.8	49.2	9	9.8
$(N_i - \tilde{N})(10^5c)$	-1.96	-1.16	-0.36	0.44	1.24	2.04	2.84	3.64	4.44
Grouping Variance $v_i(N_i - \tilde{N})^2$	38.416	21.529	4.1472	4.6464	12.300	8.3232	48.393	13.249	19.713

$$\tilde{N} = \left(\sum_i^m v_i N_i\right)\Big/n = 5.36 \quad s = \left[\sum_i^m v_i\left(N_i - \tilde{N}\right)^2\Big/(n-1)\right]^{1/2} = 1.315$$

$P(N1 < N < N2) = P(N > N1) - P(N > N2) = 0.9641 - 0.8828 = 0.0813$, as shown in Figure 9.5.

Since there are 100 specimens, the occurrence frequency is $0.0813 \times 100 = 8.1$ (times), and similar calculations can be done using Excel's built-in functions (Table 9.6).

At the beginning of this section, it was emphasized that the frequency of each group should preferably be greater than 5. Therefore, groups 6, 7, 8, and 9 are combined, and groups 0 and 1 are combined to create a new table with m = 6:

The degrees of freedom $v = 6 - 1 - 2 = 3$, because the two parameters μ and σ of the Gaussian probability density can be replaced by Ñ and s in this case. Also, note that in this case, a smaller η value is better, so the lower limit of χ_a^2 can be set to zero (Figure 9.6).

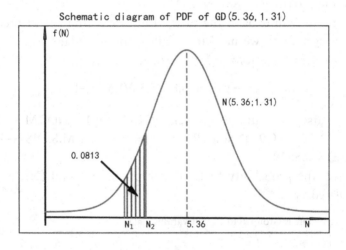

FIGURE 9.5 PDF's schematic diagram of N (5.36, 1.31).

TABLE 9.6 Data Obtained by Regrouping Fatigue Life

Group Number: i	1	2	3	4	5	6
Lower interval limit of group	3	3.8	4.6	5.4	6.2	7
Actual freqauency $v_i(c)$	10	16	32	24	8	10
Standard Fraction: u_p	−1.7974	−1.1881	−0.5788	0.03046	0.63976	1.24905
Theoretical frequency: $E_i(c)$	11.7	16.4	23.1	22.7	15.5	10.5
$v_i - E_i(c)$	−1.7	−0.4	8.9	1.3	−7.5	−0.5
$(E_i - v_i)^2/E_i$	0.24701	0.00976	3.429	0.07445	3.62903	0.02381
$\eta = \sum_i^m \left[(E_i - v_i)^2/E_i \right] =$	7.41306	$P(\chi^2 > \chi_a^2) = 10\%, \chi_a^2 =$		6.25139		

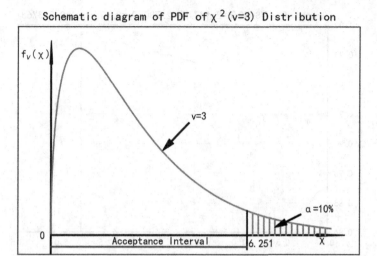

FIGURE 9.6   Schematic diagram of χ^2-distribution ($v = 3$)'s PDF and unilateral significance.

Because $\eta > \chi_a^2$, it falls outside the acceptable range; it can't be considered that the population of this sample is normal.

If it does not obey the normal distribution, does it obey the Weibull distribution? This is not only an "interesting" question but also a practical one. Use the Zhentong Gao method to solve this problem. The following results can be obtained by slightly modifying the code of the ZT Gao method in Chapter 8:

```
Nav= 5.32  s= 1.29,Nm= 5.07
km= 2780,  r= 0.993763,bm= 2.147,  lamda= 2.87,N0= 2.780
Between Gaussian dist. and ideal reliab.r=0.99051,R^2= 0.98111
Between Weibull dist. and ideal reliab.r=0.99515,R^2=   0.99032
```

From this, it can be seen that the Weibull distribution fits better in terms of reliability compared to the Gaussian distribution. In this case, there is no need for a χ^2-test to reach this conclusion. Most importantly, when the chi-square test negates that the data follows a Gaussian distribution, it does not provide information about what distribution these data follow (Figures 9.7 and 9.8).

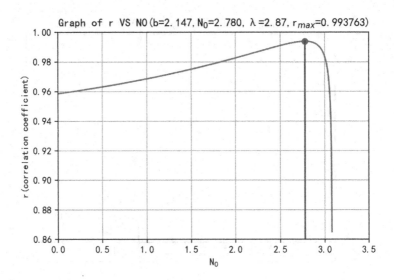

FIGURE 9.7   Chart of correlation coefficient VS N0.

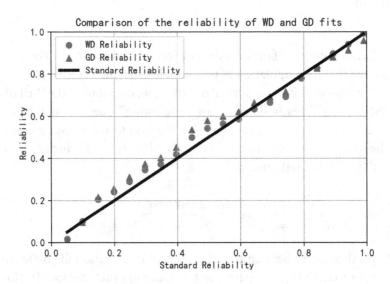

FIGURE 9.8   Comparison chart of fitting reliability between Gaussian and Weibull distribution.

9.3 t-TEST

9.3.1 t-Distribution PDF

In Chapter 8, it was mentioned that the t-test requires a large sample because the population variance is unknown. Is it possible to perform the test without knowing the population standard deviation? In 1908, the British statistician

William Sealy Gosset (1876–1937) proposed the t-test. The advantage of this method is that it does not require knowledge of the population standard deviation to perform the test. Gosset developed this test because his work did not allow for a large sample size. He worked at a brewery in Ireland, and due to cost considerations, the brewery could only provide him with a small number of samples for testing. Therefore, Gosset came up with the t-distribution.[2] Now, let's explain the origin of the t-distribution PDF.

We consider the following random variable:

$$\eta = (\chi^2/v)^{1/2} \tag{9.28}$$

Let η and χ^2 be y and x, respectively; η PDF is g(y), while χ^2 PDF $f_v(x)$ has been given in (9.7), it can be obtained that:

$$f_v(x) = [(1/2)^{v/2}/\Gamma(v/2)]x^{v/2-1}\exp(-x/2)$$

and,

$$y = (x/v)^{1/2}, \text{ inverse function } x = G(y) = vy^2 \tag{9.28a}$$

Notice again, $g(y)dy = f_v(x)dx = f_v(G(y))dG(y)$, therefore:

$$g(y) = f_v(G(y))[dG(y)/dy] \tag{9.29}$$

Obviously:

$$f_v(G(y)) = [(1/2)^{v/2}/\Gamma(v/2)](vy^2)^{v/2-1}\exp(-vy^2/2)$$

and,

$$dG(y)/dy = 2vy,$$

it can be obtained that

$$g(y) = 2[(v/2)^{v/2}/\Gamma(v/2)]y^{v-1}\exp(-vy^2/2), \quad (0 < y < \infty) \tag{9.30}$$

Let u be the standard normal random variable N(0,1); the PDF of which is:

$$\varphi(u) = [1/(2\pi)^{1/2}]\exp(-u^2/2), (-\infty < u < \infty) \tag{9.31}$$

Obviously, the random variable U and η are independent of each other, and the ratio of them is defined as:

$$t = U/\eta = U/[(\chi^2/v)^{1/2}], (-\infty < t < \infty) \tag{9.32}$$

To calculate the probability $P(t_x < t_0)$ that t_x is less than a certain value t_0, we can obtained (Gao, 1986):

$$P(t_x < t_0) = \int_0^\infty \int_{-\infty}^{t_0 y} f_v(u)g(y)dudy \tag{9.33}$$

In order to integrate, a new variable $u = zy$ is introduced, which can be obtained according to the coordinate transformation formula of double integration:

$$P(t_x < t_0) = \int_0^\infty \int_{-\infty}^{t_0 y} f_v(z,y)g(y)|\partial(u,y)/\partial(z,y)|dudy \tag{9.34}$$

where, $|(u, y)/(z, y)|$ is called the Jacobian determinant, and its value is:

$$|\partial(u,y)/\partial(z,y)| = \begin{vmatrix} \partial u/\partial z & \partial u/\partial y \\ \partial y/\partial z & \partial y/\partial y \end{vmatrix} = \begin{vmatrix} y & z \\ 0 & 1 \end{vmatrix} = y$$

So:

$$P(t_x < t_0) = \left[(2/\pi)^{1/2}\right]\left[(v/2)^{v/2}/\Gamma(v/2)\right]\int_{-\infty}^{t_0} dz \int_0^\infty y^v \exp\left[-(z^2+v)y^2/2\right]dy \tag{9.35}$$

In a further setting:

$$w = (z^2+v)y^2/2 \tag{9.36}$$

we have:

$$y = [2w/(z^2+v)]^{1/2} \rightarrow dy = dw/[2w(z^2+v)]^{1/2}$$

and,

$$P(t_x < t_0) = \pi^{-1/2}\left[v^{v/2}/\Gamma(v/2)\int_{-\infty}^{t_0}\left[(z^2+v)\right]^{-(v+1)/2}dz\int_0^\infty w^{(v+1)/2-1}\exp(-w)dw\right.$$

While noting the definition of gamma function,

$$\Gamma(\alpha)=\int_{0}^{\infty} w^{\alpha-1}\exp(-w)dw$$

Therefore:

$$P(t_x < t_0)=\pi^{-1/2}\left[v^{v/2}\Gamma\left[(v+1)/2\right]/\Gamma(v/2)\right]\int_{-\infty}^{t_0}\left[(z^2+v)\right]^{-(v+1)/2} dz$$

and,

$$P(t_x < t_0)=\int_{-\infty}^{t_0}\left\{\Gamma\left[(v+1)/2\right]\middle/\left[(\pi v)^{1/2}\Gamma(v/2)\left(1+z^2/v\right)^{(v+1)/2}\right]\right\}dz \quad (9.37)$$

We have:

$$P(t_x < t_0) = \int_{-\infty}^{t_0} h(t)dt \qquad (9.38)$$

where h(t) is given by:

$$\Gamma[(v+1)/2]/[(\pi v)^{1/2}\Gamma(v/2)](1+t^2/v)^{-(v+1)/2} \qquad (9.39)$$

$P(t_x < t_0)$ is the distribution function of t-distribution, and h(t) is its PDF.

Here notice $\Gamma(1/2) = \pi^{1/2}$, and the definition of gamma function can be obtained:

$$h(t) = 1/\{[B(1/2,v/2)v^{1/2}](1+t^2/v)^{(v+1)/2}\} \qquad (9.39a)$$

Now, let's consider the t-distribution from another perspective. The core of t-test is to replace Gaussian distribution with t-distribution, so as to solve the test problem that there are few samples and the standard deviation of the whole sample is unknown. The premise of use is that the population distribution conforms to Gaussian distribution. Gossett introduced:

$$t = (x - \mu)/[s/(n - 1)^{1/2}] \qquad (9.40)$$

And the corresponding PDF [Fitz, 1978] of t-distribution is:

$$f(t) = [\Gamma(n/2)/(n - 1)^{1/2}\Gamma(1/2)\Gamma((n - 1)/2)](t^2 + 1)^{-n/2} \qquad (9.41)$$

Many people prefer to replace n with the degree of freedom v, $v = n - 1$, i.e.:

$$f(t) = [\Gamma((v+1)/2)/v^{1/2}\Gamma(1/2)\Gamma(v/2)](t^2/v + 1)^{-(v+1)/2} \qquad (9.42)$$

Noting the definition of beta function, $b(m,n) = \Gamma(m)\Gamma(n)/\Gamma(m+n)$, so equation (9.42) can become:

$$f(t) = 1/\{[B(1/2,v/2)/v^{1/2}](t^2/v+1)^{(v+1)/2}\} \qquad (9.42a)$$

It is not difficult to find that this formula is the same as equation (9.39a). And this t-distribution is very close to the normal distribution (Fisz, 1978), especially when n→∞, it tends to normal distribution. That is, it is a very good approximation of normal distribution, which better solves the test problem that there are few samples and the standard deviation of population is unknown. In fact, $n > 30$ is quite good. This is why when the sample size is greater than 30, population distribution can be tested by t-test even if it is not normal.

9.3.2 t-Test and Examples

The principle of testing is the same, that is, first put forward a null hypothesis, assume that a positive conclusion is correct, and then start testing. If the test results support the null hypothesis, then this null hypothesis can be accepted. Otherwise, it will not be accepted, that is, the null hypothesis is "falsified".

Now, let's discuss the specific steps of the t-test. Suppose $\tilde{X}$ is the sample mean of a sample of size n from a population with distribution $N(\mu, \sigma^2)$, and the standard Gaussian variable is:

$$U = (\tilde{X} - \mu)/(\sigma/n^{1/2}) \qquad (9.43)$$

From equation (9.32), we have:

$$t_x = (\tilde{X} - \mu)/(\sigma/n^{1/2})/\left[(\chi^2/v)^{1/2}\right] = (\tilde{X} - \mu)(nv)^{1/2}/\left[\alpha(\chi^2)^{1/2}\right] \qquad (9.44)$$

Using Theorem 9.4 from Section 9.1.1, where $\chi^2 = s_x^2(n-1)/\sigma^2$ has $v = n - 1$ degrees of freedom; we substitute it into the above equation:

$$t_x = (\tilde{X} - \mu)(nv)^{1/2}/\left[s_x(n-1)^{1/2}\right] = (\tilde{X} - \mu)(n)^{1/2}/s_x \qquad (9.45)$$

Now, let's make a null hypothesis: $\mu = \mu_0$, i.e.,

$$t_x = \left(\tilde{X} - \mu_0\right)(n)^{1/2} / s_x \tag{9.46}$$

If in a single sampling we obtain the sample mean and sample standard deviation as $\tilde{X}$ and s:

$$t = \left(\tilde{X} - \mu_0\right)(n)^{1/2} / s \tag{9.47}$$

Therefore, when given significance level α, you can obtain t_α from the following equation using a table or Excel:

$$\int_{t_a}^{\infty} h(t)dt = \alpha/2 \tag{9.48}$$

$$-t_\alpha < t < t_\alpha \text{ or } |t| < t_\alpha \tag{9.49}$$

Example 9.5

Based on past experience, the average logarithmic fatigue life of a part under long-term cyclic stress is $\mu_0 = 1.8$kc. After process improvement, a batch of parts were produced, and eight parts were selected for fatigue testing. As shown in the following table, is the average logarithmic fatigue life improved (Gao, 1986)?

Solution:

We will use Excel to solve this problem (Table 9.7).

Note that t_α in the table is obtained by the formula ROUND (ABS(T.INV(0.025,7)),3). The first function ROUND is to display three decimal places; the second function ABS means absolute value, because T.INV(0.025,7) is a negative number. And T.INV(0.025,7) represents the inverse function of t-distribution with a significance of 5% and degree of freedom of 7, and it is necessary to take

TABLE 9.7 Fatigue Life Data

Fatigue Life N_i(kc)	69.2	85	87.3	89	93.5	96	107	121
$x_i = \lg(N_i)$		1.8401	1.9	1.941	1.9494	1.9708	1.9823	2.0294	2.0828
$t = (\tilde{x} - 1.8)^* n^{1/2} / s =$	6.52965	$s =$	0.0718	$\tilde{x} =$	1.96565	$v = 7, \alpha = 5\%, t_\alpha =$		2.365

a positive value for t_α. It is not difficult to see that if $t > t_\alpha$, then t is not in the acceptance range, and there is a significant difference between $\tilde{X} > \mu_0$, so the null hypothesis is not valid. But $\tilde{X} > \mu_0$; this indicates that the fatigue life has been significantly improved after the process improvement.

9.3.3 Interval Estimation of Normal Population Mean

In the previous section, we discussed point estimation. Now, we will introduce interval estimation using the t-test. Its advantage is that it does not require knowledge of the population variance. You only need to choose a confidence level γ, as shown in the shaded area in the figure below, and the upper and lower limits of the confidence interval can be determined by the following equation (Figure 9.9):

$$\int_{t_\gamma}^{\infty} h(t)dt = (1-\gamma)/2 \tag{9.50}$$

The confidence interval of t is $(-t_\gamma, t_\gamma)$, i.e.:

$$-t_\gamma < (\tilde{X} - \mu)(n)^{1/2}/s_x < t_\gamma \tag{9.51}$$

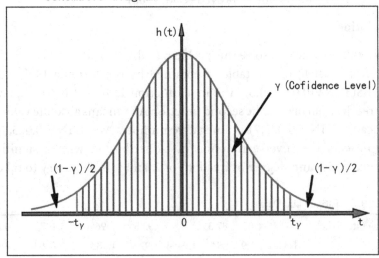

FIGURE 9.9 Schematic diagram of t-distribution's PDF and confidence.

After one sampling, $\tilde{X}$ and s_x can take the values of $\tilde{x}$ and s, respectively:

$$-t_\gamma < \left(\tilde{X}-\mu\right)(n)^{1/2}\big/s < t_\gamma \tag{9.52}$$

and,

$$\tilde{X} - st_\gamma n^{-1/2} < \mu < \tilde{X} + st_\gamma n^{-1/2} \tag{9.53}$$

Example 9.6

Still use the data in Example 9.5. So, for n = 8, the sample mean and mean square deviation are:

$$\tilde{X} = 1.966 \text{ and } s = 0.072$$

If the confidence of $\gamma = 95\%$ is taken, then as pointed out earlier, the value of t_γ is the same as the previous value of t_α, that is, both are 2.365. According to (9.53), the confidence interval of the population mean μ can be obtained:

$$1.966 - 0.072*2.365/8^{1/2} < \mu < 1.966 + 0.072*2.365/8^{1/2}$$

i.e.,

$$1.906 < \mu < 2.026$$

Also, considering: $\mu = \lg N_{50}$, we have: $10^{1.906} < N_{50} < 10^{2.026}$: so, the confidence interval for N_{50} is:

$$10^{1.906} < N_{50} < 10^{2.026}(kc)$$

and,

$$8.05*10^4 < N_{50} < 10.6*10^4(c)$$

9.3.4 t-Test of Three-Parameter Weibull Distribution and Interval Estimation of Shape Parameter

The previous discussions on the u-test, t-test, and chi-square test were all focused on the Gaussian distribution. It seems challenging to apply these tests to the Weibull distribution, which led to the introduction of the determination coefficient for curve fitting in Section 8.1. However, we still attempt to use the t-test to check if sample data follows the Weibull distribution. To do this, we employ the least squares method. In Section 8.1, we obtained the following equation:

$$Y_i = bX_i + d \tag{9.54}$$

here,

$$Y_i = \ln(\ln(1/p_i)), \ X_{i=} \ln(x_i - x_0) \tag{9.55}$$

and,

$$p_i = 1 - i/(n+1), \text{ and } d = -b\ln(\lambda); \ \lambda = \exp(-d/b) \tag{9.56}$$

The undetermined coefficient b in equation (9.54) represents the shape parameter of the Weibull distribution, and the scale parameter can also be determined by 'b' and another undetermined coefficient d. So, estimating b and d allows us to estimate the Weibull distribution's parameters. However, in Section 8.1, the significance and confidence interval estimation of these parameters were not provided. Here, we address this issue, even though the theoretical derivation is somewhat complex (Xu & Gao, 2022).

A reasonable assumption is that the regression errors u_i follow a $N(0, \sigma^2)$ distribution:

$$u_i = Y_i - (bX_i + d) \tag{9.57}$$

Now, let's discuss the estimation of b, d, and σ^2. First, we need to prove that the estimates obtained by the least squares method, $\hat{b}$ and $\hat{d}$, are the best linear unbiased estimators. Using the least squares method, we get:

$$\hat{b} = \Sigma(Y_i - \tilde{Y})(X_i - \tilde{X})/\Sigma(X_i - \tilde{X}) \tag{9.58}$$

$$\hat{d} = \tilde{Y} - \hat{b}\tilde{X} \tag{9.59}$$

here,

$$\tilde{X} = \Sigma X_i/n; \ \tilde{Y} = \Sigma Y_i/n \tag{9.60}$$

So:

$$\hat{b} = \Sigma Y_i \left(X_i - \tilde{X}\right) \Big/ \Sigma \left(X_i - \tilde{X}\right)^2 - \Sigma \left(X_i - \tilde{X}\right) \Big/ \Sigma \left(X_i - \tilde{X}\right)^2$$

$$\hat{b} = \Sigma f_i Y_i \tag{9.61}$$

where,

$$f_i = (X_i - \tilde{X}) / \Sigma (X_i - \tilde{X})^2, \text{ and } \Sigma(X_i - \tilde{X}) = 0 \tag{9.62}$$

This proves the linearity of $\hat{b}$. It's not hard to be proven:

$$\Sigma f_i = 0; \ \Sigma f_i(X_i - \tilde{X}) = \Sigma f_i X_i = 1 \tag{9.62a}$$

Because equation (9.59) is easy to prove that: $\hat{d}$ is also a linear function of Y_i:

$$\hat{d} = \Sigma g_i Y_i \tag{9.63}$$

where,

$$g_i = \left(1/n - \tilde{X} f_i\right) \tag{9.63a}$$

Using equations (9.57 and 9.61), we have:

$$\hat{b} = \Sigma f_i (bX_i + d + u_i) = b\Sigma f_i X_i + d\Sigma f_i + \Sigma f_i u_i = b + \Sigma f_i u_i \tag{9.64}$$

Here, notice again equations (9.62 and 9.62a). From equation (9.57), we also have:

$$\hat{d} = \Sigma g_i (bX_i + d + u_i) = b\Sigma g_i X_i + d\Sigma g_i + \Sigma g_i u_i = d + \Sigma g_i u_i \tag{9.65}$$

Notice that (9.63a):

$$\Sigma g_i = \Sigma(1/n - \tilde{X} f_i) = 1, \ \Sigma g_i X_i = \Sigma X_i (1/n - \tilde{X} f_i) = 0$$

Therefore, as to prove the unbiasedness of parameters $\hat{b}$ and $\hat{d}$, i.e.:

$$E(\hat{b}) = b; \ E(\hat{d}) = d \tag{9.66}$$

In fact, from equation (9.64) we can get: $E(\hat{b}) = E(b + \Sigma f_i u_i) = b + \Sigma E(f_i)$ $E(u_i) = b$.

We can also prove: $E(\hat{d}) = d$.

Now, let's further find the variance of $\hat{b}$:

$$Var(\hat{b}) = Var(b + \Sigma f_i u_i) = Var(b) + Var(\Sigma f_i u_i)$$

$$= \Sigma Var(f_i) Var(u_i) = \sigma^2 / \Sigma(X_i - \tilde{X})^2 \qquad (9.67)$$

and,

$$Var(\hat{d}) = Var(d) + Var(\Sigma g_i u_i) = \sigma^2 E(X^2) / \Sigma(X_i - \tilde{X})^2 \qquad (9.67a)$$

Here, notice that:

$$\Sigma g_i^2 = \Sigma(1/n - \tilde{X} f_i)^2 = \Sigma\left[1/n^2 - 2\tilde{X} f_i/n + (\tilde{X} f_i)^2\right]$$

$$= 1/n + \tilde{X}^2/\Sigma(X_i - \tilde{X})^2 = E(X^2)/\Sigma(X_i - \tilde{X})^2$$

Since σ^2 is unknown, we need to estimate it from the observed data. The observed error is:

$$\hat{\varepsilon}_i = Y_i - \tilde{Y}_I = (Y_i - \hat{Y}) - (\hat{Y}_i - \hat{Y}) \qquad (9.68)$$

where Y_I is determined by equation (9.57), and we can obtain:

$$\tilde{Y} = b\tilde{X} + d + \tilde{u}, \text{ and } \hat{Y}_i = \left(\hat{b} X_i + \hat{d}\right)$$

Therefore,

$$Y_i - \tilde{Y} = b(X_i - \tilde{X}) + u_i - \tilde{u}; \quad \hat{Y}_i - \hat{Y} = \hat{b}(X_i - \tilde{X}) \qquad (9.69)$$

Equation (9.68) can become:

$$\hat{\varepsilon}_i = (b - \hat{b})(X_i - \tilde{X}) + (u_i - \tilde{u}) \qquad (9.70)$$

Thus,

$$E(\Sigma \hat{\varepsilon}_i^2) = E\left[(b - \hat{b})^2 \Sigma(X_i - \tilde{X})^2 + 2(b - \hat{b})\Sigma(X_i - \tilde{X})(u_i - \tilde{u}) + \Sigma(u_i - \tilde{u})^2\right]$$

Notice that:

$$E\left[\left(\hat{b}-b\right)^2\right]=E\left[\left(\hat{b}-E(\hat{b})\right)^2\right]=Var(\hat{b})=\sigma^2\Big/\Sigma\left(X_i-\tilde{X}\right)^2$$

i.e.:

$$E\left[\left(b-\hat{b}\right)^2\Sigma\left(X_i-\tilde{X}\right)^2\right]=\sigma^2$$

From equation (9.64), it is available:

$$E\left[(b-\hat{b})\Sigma\left(X_i-\hat{X}\right)(u_i-\tilde{u})\right]=-E\left[\Sigma f_i u_i\Sigma(X_i-\tilde{X})u_i\right]=-\sigma^2$$

Here, notice that:

$$\Sigma\left(X_i-\tilde{X}\right)\tilde{u}=0,\quad \Sigma f_i\left(X_i-\tilde{X}\right)=1,\quad \Sigma u_i^2=\Sigma(u_i-0)^2=\sigma^2$$

At the same time because $E\left[\Sigma(u_i-\tilde{u})^2\right]=(n-1)\sigma^2$, we can obtain:

$$E\left(\Sigma\hat{\varepsilon}_i^2\right)=\sigma^2-2\sigma^2+(n-1)\sigma^2=(n-2)\sigma^2 \qquad (9.71)$$

Therefore, unbiased estimation of σ^2 is that:

$$\hat{\sigma}^2=\Sigma\left(Y_i-\hat{Y}_i\right)^2\Big/(n-2) \qquad (9.72)$$

So, whether the Weibull distribution is significant under the condition of confidence $\gamma=1-\alpha$ (significance) and the estimation of the Weibull parameter itself. Start with shape parameter b. Notice that the estimated value $\hat{b}$ of b obeys Gaussian distribution N $(b, \sigma^2/\Sigma(X_i-\tilde{X})^2)$, where σ^2 can be replaced by the estimated unbiased: $\hat{\sigma}^2=\Sigma\left(Y_i-\hat{Y}_i\right)^2\Big/(n-2)$.

$$\text{"Standardize" it}: t=(\hat{b}-b)\Big/\hat{\sigma}_b \sim t(n-2) \qquad (9.73)$$

where $\hat{\sigma}_b^2=\hat{\sigma}^2\Big/\Sigma\left(X_i-\tilde{X}\right)^2$, because the freedom of $\hat{\sigma}^2$ is n – 2. So, the freedom of $\hat{\sigma}_b^2$ is n – 2 too; therefore, t obeys the t(n – 2) distribution. We can do the following tests:

Null hypothesis is that b = 0 (it means that at the significance level α taking Weibull distribution is incorrect).

Alternative hypothesis is that $b \neq 0$ (it means that at the significance level α taking Weibull distribution is correct).

If $|t| \leq t_{\alpha/2}(n-2)$, accept the null hypothesis; otherwise, accept the alternative hypothesis   (9.3-47).

As for the confidence interval of b, it is easier to determine. At this time, $1 - \alpha$ is used to represent the confidence level, and then using equation (9.73), the confidence interval of b can be obtained, which will be illustrated with specific examples below:

Example 9.7

Using the data from Example 8.2, and with some modifications to the Python code by using the Zhentong Gao method (Xu & Gao, 2022), the following results are obtained:

```
N= [350, 380, 400, 430, 450, 470, 480, 500, 520, 540, 550, 570, 600, 610, 630, 650, 670, 730, 770, 840]
N0= 276.6, r= 0.99922, b= 2.040, lamda= 320.98, alpha= 0.05
t= 25.26 >t_alp= 2.10, so the population distribution conforms to Weibull distribution.
Confidence interval of population shape parameter b: ( 2.000, 2.080 )
Confidence interval of parameter d: ( -11.993, -11.550 )
Confidence interval of population scale parameter lamda: ( 319.5, 322.5 )
```

It can be seen that it is feasible to determine the confidence interval of b and d by this method, but the confidence interval of safe life cannot be determined. This method can test whether the sample conforms to Weibull distribution under a certain confidence level. However, the confidence interval of three parameters of Weibull distribution cannot be determined at the same time, and the confidence interval of fatigue life under a certain reliability cannot be determined.

9.3.5 Confidence Interval Curve of Fatigue Life of Three-Parameter Weibull Distribution

To address the problem of confidence interval of fatigue life under a certain reliability, the concept of rank (order) distribution statistic is still needed (Fu et al., 1992). Actually, let $x_1, \ldots, x_n$ be an independent random variable from the same population, and let $x_1 < x_2 < \ldots < x_n$, and let P_k be less than the frequency value of the occurrence of the kth X_i, then it is easy to get the PDF of the kth random variable as:

$$f_n(P_k) = [n!/(k-1)!(n-k)!]P_k^{k-1}(1-P_k)^{n-k} \qquad (9.74)$$

This leads to the conclusion:

$$f_n(x) \sim Be(k, n-k+1) = x^{k-1}(1-x)^{n-k}/B(k, n-k+1) \qquad (9.75)$$

It's worth noting that the PDF of the Beta distribution is: $Be(a, b) = x^{a-1}(1-x)^{b-1}/B(a, b)$.

To obtain a confidence interval of P_k with confidence γ, according to Fu et al. (1992), we define P_{uk} and P_{Lk} as the upper and lower confidence limits of P_k, respectively, as follows:

$$\int_0^{P_{uk}} f_n(P_k)dP_k = \gamma, \quad \int_{P_{Lk}}^1 f_n(P_k)dP_k = \gamma \qquad (9.76)$$

The confidence interval of Pk is:

$$[P_{Lk}, P_{uk}], \text{ with } P(P_{Lk} \le P_k \le P_{uk}) = 2\gamma - 1 \qquad (9.77)$$

where, $\gamma \ge 50\%$.

From (9.76), we get:

$$\int_0^{P_{u(n-k+1)}} f_n(P_{n-k+1})dP_{n-k+1} = \gamma \qquad (9.78)$$

Considering a variable substitution, $y = 1 - x$, we can write:

$$-\int_1^{1-P_{u(n-k+1)}} f_n(P_k)dP_k = \gamma \text{ and } \int_{1-P_{u(n-k+1)}}^1 f_n(P_k)dP_k = \gamma$$

By comparing this equation with the second part of equation (9.76), we find:

$$P_{Lk} = 1 - P_{u(n-k+1)} \qquad (9.79)$$

With this formula, the calculation workload can be saved by half, that is, people only need to calculate P_u. The problem is that it is not enough to calculate P_u and list it in a table for people to refer to (Fu et al., 1992). After all, it is still troublesome, and it is not easy to calculate the corresponding parameters of Weibull distribution. This is probably the reason why this method has not been well popularized and applied. Now that the computer hardware and software are so developed, there is no need to look up the table to calculate and use these values, and it can be very intelligent to "give according to needs". Very convenient for people to use. Of course, the algorithm is based on the ZT Gao method.

Example 9.8

Using the data from Example 9.7 and modifying Python code (as found in Appendix A of the book), the following results are obtained (Xu & Gao, 2022):

```
N= [350, 380, 400, 430, 450, 470, 480, 500, 520, 540, 550, 570, 600, 610, 630, 650,
670, 730, 770, 840]
In Confidence= 0.95 and Size= 20 : super limit of Confidence:
13.91,21.61,28.26,34.37,40.10,45.56,50.78,55.80,60.64,65.31,69.80,74.13,78.29,82.27,
86.04,89.59,92.86,95.78,98.19,99.74,
In Confidence= 0.95,and Size= 20 : lower limit of Confidence:
0.26,1.81,4.22,7.14,10.41,13.96,17.73,21.71,25.87,30.20,34.69,39.36,44.20,49.22,54.44,
59.90,65.63,71.74,78.39,86.09,
Parameters by average rank:b= 2.040,λ= 320.98,N0= 276.60,r= 0.99922
Parameters by super rank:b= 2.797,λ= 368.96,N0= 158.80,r= 0.99897
Parameters by lower rank:b= 1.986,λ= 343.26,N0= 333.06,r= 0.99926
Confidence Interval of b= 2.040 is:[ 1.986, 2.797 ]
Confidence Interval of N0= 276.600 is:[ 158.800, 333.060 ]
In Reliability= 0.999 :Confidence Interval of fatigue Life(287.459 )
is[190.019,343.658 ]
```

The results obtained here are almost consistent with those in Fu et al. (1992). We have improved their work, making it more intelligent and presenting the confidence curves for fatigue life. It is very convenient to use and less prone to errors (Figure 9.10).

By simply changing γ from 0.95 to 0.9, you can immediately obtain the following results:

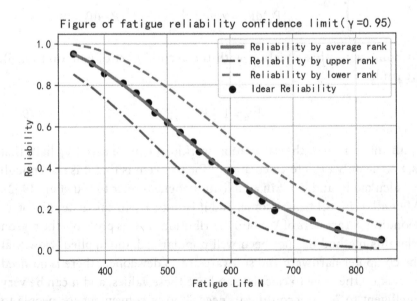

FIGURE 9.10   Curve chart for confidence limit of fatigue life with confidence of 0.95.

```
N= [350, 380, 400, 430, 450, 470, 480, 500, 520, 540, 550, 570, 600, 610, 630, 650,
670, 730, 770, 840]
In Confidence= 0.9 and Size= 20 : super limit of Confidence:
10.87,18.10,24.48,30.42,36.07,41.49,46.73,51.80,56.73,61.52,66.18,70.71,75.09,79.33,8
3.41,87.31,90.98,94.36,97.31,99.47,
In Confidence= 0.9,and Size= 20 : lower limit of Confidence:
0.53,2.69,5.64,9.02,12.69,16.59,20.67,24.91,29.29,33.82,38.48,43.27,48.20,53.27,58.51,
63.93,69.58,75.52,81.90,89.13,
Parameters by average rank:b= 2.040,λ= 320.98,N0= 276.60,r= 0.99922
Parameters by super rank:b= 2.513,λ= 336.90,N0= 204.20,r= 0.99905
Parameters by lower rank:b= 1.946,λ= 329.56,N0= 327.74,r= 0.99931
Confidence Interval of b= 2.040 is:[ 1.946, 2.513 ]
Confidence Interval of N0= 276.600 is:[ 204.200, 327.740 ]
In Reliability= 0.999 :Confidence Interval of fatigue Life(287.459 )
is[225.777,337.219 ]
```

Comparing with the result of $\gamma = 0.95$, we can find that the confidence level is reduced, so the confidence interval will be "shortened", which seems to narrow the target, but the hit rate (confidence level) is also small, so it can be said that "the fish and the bear's paw cannot have both", so we must weigh the pros and cons. This is the reason why people generally prefer to choose a confidence level of 0.95 (Figure 9.11).

There are several issues that need to be emphasized here:

1. The confidence limits mentioned above are for CDF, which are the opposite of the confidence limits for reliability, which is why the upper and lower confidence limits for CDF become the lower and upper confidence limits for reliability, respectively.

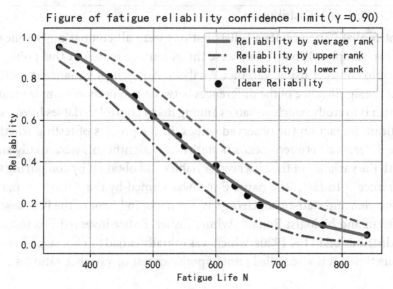

FIGURE 9.11 Curve chart for confidence limit of fatigue life with confidence of 0.9.

2. Consequently, the confidence interval for the safe life parameter N_0 is also consistent with the reliability, i.e., it needs to be "reversed". However, the confidence interval of the shape parameter b is negatively correlated with the position parameter, so it is "consistent" with the confidence limit of CDF instead of the reliability. It is easy to see that the confidence interval for b obtained in Section 9.4 is shorter, i.e., better, than the one obtained in this section under the same confidence level. The reason for this is probably that the former is calculated directly, while the latter is obtained indirectly.

3. Another issue to note is that scale parameters are not provided here λ. The confidence interval is determined because the determination of the three parameters of the Weibull distribution is correlated. First, the position parameter N_0 is determined by maximizing the correlation coefficient, then the shape parameter b is determined by the least squares method, and finally, the scale parameter is indirectly calculated λ. Calculated from the average rank $\lambda = 320.98$, surprisingly not calculated from the upper and lower limits of rank λ in the interval [329.56336.9]. Therefore, the so-called confidence interval obtained by this method is meaningless.

9.4 F-TEST

9.4.1 PDF of F-Distribution

Notice that "The previous t-distributions were all comparing the means of two samples. What if they were three, four, or more? At this point, we need to consider using Variance Analysis. Analysis of variance is used to test the significance of the differences between two or more sample means, which is to study which variables among many control variables have a significant impact on the observed values. In the process of testing whether the difference between means is statistically significant, we actually start with the variance of the observed variables and obtain it by comparing the variance".[3] In fact, "the name F-test was named by the American mathematician and statistician George W. Snedecor in honor of the British statistician and biologist Ronald Aylmer Fisher. Fisher invented this test and F-distribution in the 1920s, which was initially called the Variance Ratio".[4] Sometimes, it is also called joint hypotheses test or variance ratio test.

Now, let us first derive the so-called F-distribution of PDF. For this purpose, we can set two independent χ^2 variable X^2 and Y^2, their PDFs are f (x) and g (y), and the degrees of freedom are v, w., and then define the random variable:

$$F^* = X^2/Y^2 \tag{9.80}$$

The PDF derived from F-distribution is basically similar to the PDF derived from t-distribution in Section 9.3.1:

$$P\left(F^* < F_0^*\right) = \int_0^\infty \int_0^{F_0^* y} f(x)g(y)dxdy \tag{9.81}$$

By making a variable substitution, $x = zy$, and using the double integral transformation formula as in Section 9.1, we obtain:

$$P\left(F^* < F_0^*\right) = \int_0^\infty \int_0^{F_0^*} f(z,y)g(y)|\partial(x,y)/\partial(z,y)|dxdy \tag{9.82}$$

It can be found that $|\partial(x, y)/\partial(z, y)| = y$, and as noted in equation (9.34):

$$f_v(x) = [(1/2)^{v/2}/\Gamma(v/2)]x^{v/2-1}\exp(-x/2);$$

So, we have:

$$f(z,y) = [(1/2)^{v/2}/\Gamma(v/2)](zy)^{v/2-1}\exp(-zy/2);$$

and,

$$g(y) = [(1/2)^{w/2}/\Gamma(w/2)]y^{w/2-1}\exp(-y/2)$$

Therefore,

$$P\left(F^* < F_0^*\right) = K\int_0^{F_0^*} z^{v/2-1}dz\int_0^\infty y^{(v+w)/2-1}\exp\left[-(z+1)y/2\right]dy \tag{9.83}$$

where

$$K = [(1/2)^{(v+w)/2}/\Gamma(v/2)\Gamma(w/2)].$$

Now, introduce a new variable substitution: $h = (z + 1)y/2 \rightarrow y = [2/(z + 1)]$ h, $dy = [2/(z + 1)]dh$. We have:

$$P\left(F^* < F_0^*\right) = K \int_0^{F_0^*} \left[2/(z+1)\right]^{(v+w)/2} z^{v/2-1} dz \int_0^\infty h^{(v+w)/2-1} \exp(-h) dh$$

Noting the definition of gamma function, the rightmost integral of the above formula $= \Gamma[(v+w)/2)]$; therefore:

$$P\left(F^* < F_0^*\right) = \left\{\Gamma[(v+w)/2]/\Gamma(v/2)\Gamma(w/2)\right\} \int_0^{F_0^*} \left[1/(z+1)\right]^{(v+w)/2} z^{v/2-1} dz$$

$$(9.84)$$

Further noting the definition of the beta function: $B(m, n) = \Gamma(n/2)\Gamma(m/2)/\Gamma[(n + m)/2)]$, we get:

$$P\left(F^* < F_0^*\right) = \left[1/B(v/2, w/2)\right] \int_0^{F_0^*} \left[1/(z+1)\right]^{(v+w)/2} z^{v/2-1} dz \quad (9.85)$$

Further noting the definition of the beta function PDF of F^* is:

$$p^* = [1/B(v/2,w/2)][1/(z + 1)]^{(v + w)/2} z^{v/2 - 1} dz \quad (9.86)$$

However, in practical applications, it is often more convenient to use the following variable:

$$F' = (X^2/v)/(Y^2/w) \quad (9.87)$$

which implies:

$$F' = (X^2/v)/(Y^2/w) = (wX^2)/(vY^2) = (w/v)F^* \quad (9.88)$$

So, when F' takes the value F, F^* takes the value z:

$$F = (w/v)z \quad (9.89)$$

Resulting in:

$$z = (v/w)F, \quad dz = (v/w)dF \qquad (9.90)$$

Let p(F) be the PDF of F, and following the method used previously, we have:

$$p(F)dF = p^*(z)dz = [1/B(v/2,w/2)][1/(z+1)]^{(v+w)/2}z^{v/2-1}dz$$

Notice that (9.4-9), the above formula becomes:

$$p(F) = [1/B(v/2,w/2)][1/(vF/w+1)]^{(v+w)/2}(v/w)^{v/2}F^{v/2-1}$$

Thus, the PDF of F is given by:

$$p(F) = [v^{v/2}w^{w/2}/B(v/2,w/2)][1/(vF+w)]^{(v+w)/2}F^{v/2-1} \qquad (9.91)$$

The PDF p(F) has two degrees of freedom, v and w, sometimes referred to as the numerator and denominator degrees of freedom. It is essential to note that in equation (9.89), v and w correspond to the denominator and numerator, respectively. Therefore, using the first and second degrees of freedom in Excel is more scientific and avoids unnecessary confusion. Once these two degrees of freedom are given, the probability that F_x exceeds a certain value F_a can be calculated as follows (Figure 9.12):

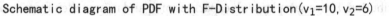

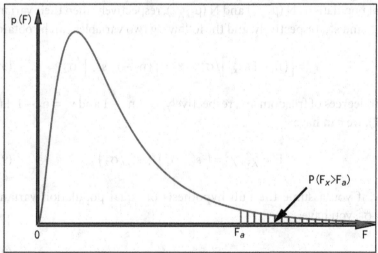

FIGURE 9.12 Schematic diagram of F-distribution ($v_1 = 10$, $v_2 = 6$)'s PDF.

$$P(F_x > F_a) = \int_{F_a}^{\infty} p(F)dF \tag{9.92}$$

There is no problem to look up the value of F_a when the numerator denominator degree of freedom and significance α are given. For example Gao (1986) points out that $F_\alpha = 4.06$ when the numerator degree of freedom and denominator degree of freedom are 10 and 6, respectively, and the significance is 5%. Of course, it's more convenient to use Excel:

F.INV(0.95,6,10)=	4.059963

Here is a brief explanation of the usage of the function F.INV; 0.95 represents the CDF of F, so when the significance is 5%, it is necessary to take $1 - 5\% = 95\%$. At the same time, pay special attention to the order of degrees of freedom, that is, the so-called "denominator", that is, the degree of freedom within a group, should be filled in first, and then the "numerator", that is, the degree of freedom between groups. If the order is reversed, the value will be different. In fact:

F.INV(0.95,10,6)=	3.217175

9.4.2 F-Test and Example

Because the PDF of F-distribution is derived from the distribution theory of χ^2 variable, the population distribution is required to obey the normal distribution. If two samples with capacities of n_1 and n_2 are taken from two different populations $N(\mu_1, \chi_1^2)$ and $N(\mu_2, \chi_2^2)$, respectively, then their variances are $s_{x_1}^2$ and $s_{x_2}^2$, respectively, and the following two variables can be obtained:

$$\chi_1^2 = \left[(n_1 - 1)\, s_{x_1}^2\right]/\sigma_1^2, \quad \chi_2^2 = \left[(n_2 - 1)\, s_{x_2}^2\right]/\sigma_2^2 \tag{9.93}$$

Their degrees of freedom are, respectively, $v_1 = n_1 - 1$ and $v_2 = n_2 - 1$. From (9.87), we can have:

$$F' = \chi_1^2/\chi_2^2 = \left(s_{x_1}^2/\sigma_1^2\right)/\left(s_{x_2}^2/\sigma_2^2\right) \tag{9.94}$$

Now, if you assume the null hypothesis of equal population variances: $\sigma_1^2 = \sigma_2^2$, you have:

$$F' = s_{x_1}^2/s_{x_2}^2 \tag{9.95}$$

If $s_{x_1}^2$, $s_{x_2}^2$ can get the value of s_1^2 and s_2^2, respectively, then you have:

$$F = s_1^2 / s_2^2 \qquad (9.96)$$

As shown in Figure 9.13, the acceptance interval (a, b) can be determined after the significance α is given. If the calculated F value is in the acceptance range, the null hypothesis can be accepted; otherwise, it will be rejected. However, in the actual situation, it is enough to find the α value for convenience (it is also more convenient to use Excel). For this reason, the greater of s_1^2 and s_2^2 is generally regarded as the numerator, but the corresponding degree of freedom is the "denominator degree of freedom" (that is, it should be placed at the back), which is the "second degree of freedom" in Excel.

Example 9.9

To study the effect of processing methods on the variance of fatigue life, assuming that the fatigue lives follow a Gaussian distribution. Two batches of parts were processed using different methods, and samples of 10 and 8 pieces were taken for fatigue life testing. The results are shown in Table 9.8. Are there significant differences in the variances of the two (Gao, 1986)?

This is completely calculated by Excel, but it should be noted that the standard deviation is given by Excel, so the variance needs to

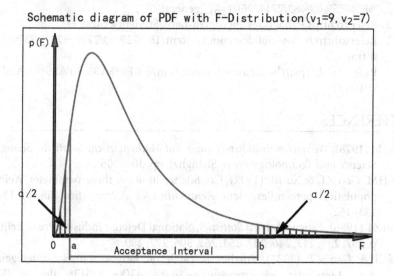

FIGURE 9.13 Schematic diagram of F-distribution ($v_1 = 9$, $v_2 = 7$)'s PDF and confidence.

TABLE 9.8 Date of Example 9.9

First Log. Fatigue Life	4.532	4.632	4.632	4.643	4.648	4.653	4.672	4.69	4.69	4.724
Sec. Log. Fatigue Life	4.839	4.924	4.945	4.954	4.973	4.978	5.033	5.083	□	□

$s_1 =$ 0.05139 $s_1^2 =$ 0.00264 $a = 5\%$, $F_\alpha = 4.19705$

$s_2 =$ 0.07255 $s_2^2 =$ 0.00526 $F = s_2^2/s_1^2 = 1.99328$

be squared. The other one has a larger variance, so it is a numerator (corresponding to the first degree of freedom). In addition, we should pay attention to $\alpha = 5\% \to \alpha/2 = 2.5\%$ when using the formula in Excel. Use the formula F.INV to calculate as follows:

$$1 - 2.5\% = 0.975 \to F_\alpha = F.INV(0.975, 7, 9) = 4.197 > 1.99 = F$$

It can be seen that the population mean square deviation of the two processing methods is statistically "equal" when the significance $\alpha = 5\%$, that is, the difference between the processing methods and the fatigue life is "insignificant".

NOTES

1 Refer to https://baike.baidu.com/item/%E5%8D%A1%E6%96%B9%E5%88%86%E5%B8%83/2714796?fr=kg_general
2 Refer to https://zh.wikipedia.org/zh-hans/Student's t-test
3 Refer to http://www.360doc.com/content/18/0728/10/27499672_773795115.shtml
4 Refer to https://baike.baidu.com/item/F%E6%A3%80%E9%AA%8C/9910842

REFERENCES

Fisz M (1978), *Wahrscheinlichkitsrechnung und Mathematische Stastistik*, Shanghai Science and Technology Press, Shanghai, pp. 304–306.

Fu HM, Gao ZT, & Xu RP (1992), Confidence limits of three-parameter Weibull population percentiles, *Acta Aeronautica et Astronautica Sinica*, 13(3): 153–162.

Gao ZT (1986), *Fatigue Applied Statistics*, National Defense Industry Press, Beijing, p. 237, 239, 243, 246, 247, 253, 261, 306, 277, 280.

Xu JJ & Gao ZT (2022), Further research on fatigue statistics intelligence, *Acta Aeronautica et Astronautica Sinica*, 43(8): 225138. doi: 10.7527/S1000-6893.2021.25138

CHAPTER 10

Application of Digital-Experiments to the Study of Weibull Distribution

10.1 DIGITAL-EXPERIMENTS AND MACHINE LEARNING

As we all know, modern electronic computers fundamentally use physical components, such as early electronic (diode, triode) tubes, followed by semiconductor transistors, integrated circuits, and now chips, to simulate human computational processes. Although they are called digital computers, they essentially simulate human calculations and various cognitive activities. In fact, the earliest computers were analog machines, but nowadays, analog computers have almost disappeared.

Why is that? It's because, just like paintings created by artists are "analog", the graphics produced by computers are "digital". The difference lies in the fact that the former is continuous and natural, while the latter is discrete and sometimes doesn't seem as "natural". Of course, this is just a metaphor, but the difference is substantial. For example, before the widespread use of computers in the 1980s, creating an animation required countless paintings and a lot of time. But now, making animations through computers saves an incredible amount of effort because computers have digitized the artwork.

DOI: 10.1201/9781003488477-13

Therefore, when we talk about "simulation" today, it's not the original sense of simulation but rather digital simulation. In the 1960s, students and faculty at Beihang University (BUAA) took pride in BUAA's wind tunnel because there were very few units in China at the time that had wind tunnels. The original data for aircraft design came from these wind tunnels... However, conducting a wind tunnel test consumes an enormous amount of electricity. Now, we have digital wind tunnels (Wang, 1988). Before formally using a physical wind tunnel, you can first "blow" in the digital wind tunnel. We did not know how many resources this could save. Of course, such wind tunnels are also applicable to cars, high-speed trains, and even tall buildings. Similar digital simulations (and further Digital-Experiments) have become indispensable in various fields. Here, it's essential to emphasize that the digital simulation referred to here means using digital technology to simulate existing things, while Digital-Experiments mean using digital technology to simulate assumed events. Digital simulation is the foundation of Digital-Experiments, and Digital-Experiments are the inevitable outcome of digital simulation's development.

Unfortunately, when searching online, "Digital-Experiments" typically refer to logic circuit experiments, which have a completely different meaning from the definition provided here. There are also books like "Mathematical Experiments" (Li, 2004). Nevertheless, "mathematical experiments" usually only refer to using computers to solve mathematical problems, but the significance of digital simulation or Digital-Experiments mentioned here is much broader.

Similarly, for "fatigue statistics", digital simulation is either essential or indispensable (Xu, 2022). Theoretically, the Weibull distribution was daunting because of its complex mathematical relationships, but with computers, it can be handled effortlessly. You can perform various Digital-Experiments almost as you wish, which is the main topic of discussion below. In terms of practical applications, fatigue testing is very time-consuming and labor-intensive. Can it be digitized like wind tunnels? The challenge seems not to be in computers but in the fact that people still don't have a deep understanding of the fatigue mechanism and haven't found an equation for fatigue similar to fluid dynamics. Of course, such an equation may not exist for fatigue, but through "machine learning" (Bowles, 2017), it is still possible to help people "create" a "digital fatigue machine". To achieve this, a thorough understanding of WD's digital simulation and experiments is essential, which is the primary content to be discussed in this chapter. This chapter specifically introduces a highly effective Digital-Experimental method—the Bootstrap (Montgomery et al., 2019; and Efron & Hastie, 2019), which can only be realized through computers—and uses this method to determine the confidence intervals of the three parameters of Weibull distribution.

10.2 DIGITAL-EXPERIMENT ON WEIBULL DISTRIBUTION

10.2.1 Random Generator of Three-Parameter Weibull Distribution

To perform a Digital-Experiment on the three-parameter Weibull distribution, a basic condition is to generate a variety of sample data at will, that is, to construct a random "generator" of the three-parameter Weibull distribution. There are no three-parameter Weibull distribution random generators in general software libraries, for the simple reason that it is "natural" for most users to treat the location parameter x_0 as zero, from a mathematical perspective. It seems that it only reflects the translation of one coordinate axis and does not seem to affect other properties. However, as pointed out in Section 3.2, the location parameter in Weibull distribution is called "safe life" in fatigue statistics, which is generally not zero, so it is unrealistic to assume zero, and it is also not advisable. It cannot and should not be set to zero for serious study of the Weibull distribution. According to this principle, the "Random Generator of Three-Parameter Weibull Distribution" is composed of the following code in Python:

```
def randomWN(NN,b,λ,x0,G):
    np.random.seed(G)
    NR=np.random.rand(NN)
    N=x0+λ*np.exp(np.log(np.log(1/NR))/b)
    N.sort()
    return N
```

Among them, NN is the capacity (size) of the sample to be generated, b, λ, x_0 are the shape parameters, scale parameters, and location parameter (in the study of fatigue life, it is generally expressed by N_0, which is called the safe life parameter) of the Weibull distribution, respectively. And G represents the "seed" for generating random numbers, so that the generated samples can be repeated, which is very important for Digital-Experiments. The following section is to change the samples of different Weibull distributions by altering different parameters to test the pros and cons of the ZT Gao method, the analytical method, and Gaussian distribution fitting.

10.2.2 Digital-Experiment on Various Parameters
Related to Weibull Distribution

Similar to actual experiments, in Digital-Experiments, only one parameter is changed while keeping other parameters constant. This approach helps in understanding the effect of each parameter. The code for this section can be found in the postscript Code B at the end of the book.

10.2.2.1 Experiment 1: Change only the Location x_0

Changing only x_0 to 0, 1, and 5, while keeping other factors NN = 50, $G = 1$, $b = 2.5$, and $\lambda = 5$ constant, using Python code, we obtain the following results:

It can be seen that the shape factor $b = 2.186$ obtained by the ZT Gao method is within its confidence interval [1.943, 2.217], and this confidence interval does not include the true population shape parameter $b = 2.5$, and the safe life $x_0 = 0$, also within its confidence interval [0, 0.59], also includes the population safe life $x_0 = 0$. $\lambda = 5.37$ is also within the confidence interval [4.51, 5.55]. In contrast, the solution obtained by the analytical method deviates significantly from the true value of the population parameters. Although the difference is very small both from the correlation coefficient and the determination coefficient compared to the ZT Gao method, this also shows that it is possible to make errors in the discrimination from the coefficients of determination alone, especially if the plot is fitted alone (Figure 10.1 and Table 10.1).

In the case of $x_0 = 1$, the shape factor $b = 2.253$ obtained by the ZT Gao method is within its confidence interval [2.217, 2.558], and this confidence interval also includes the true population shape parameter $b = 2.5$. Similarly, the safe life $x_0 = 0.91$ is also within its confidence interval [0, 1.59] and includes the population safe life $x_0 = 0$. Obviously, the scale parameter is not so right; this is also consistent with the conclusions of Section 9.3.5. It is worth noting that the solution obtained by the analytical method is indeed the same as when $x_0 = 0$. Why? It is true that for the analytical method, the change in x_0 does not affect the median, mean, and mean square of the data, so the result is the same, and the same is true for the GD. Here, why did the use of the ZT Gao change? This is because when $x_0 = 0$ is taken, the change interval is [0,0], but now, it is changed to $x_0 = 1$. Then, the change interval is [0,1], and the location of the maximum correlation coefficient obtained is no longer zero, so the result is different.

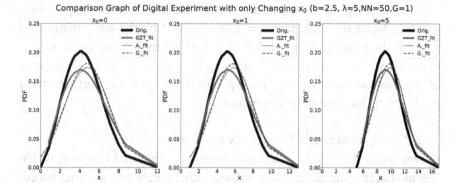

FIGURE 10.1 Comparison plot of three different fits when only changing the location parameter x_0 when taking random samples from the Weibull distribution. Note: In the legend, "orig". is an abbreviation for "original", representing sample data; "GZT_fit" means the fitting curve of the PDF of WD obtained by the ZT Gao method; "A_fit" represents the fitted curve of the PDF of WD obtained by the analytical method. "G_fit" represents the curve obtained by fitting the sample data with the PDF of GD. And, subsequent *figures also apply the same abbreviations defined for this figure below in this chapter. The legends in Figures 10.2–10.5 for the remainder of this chapter are the same as this legend.

TABLE 10.1 Comparison of Three Different Fits for Weibull Distribution Random Samples with Different Location Parameter x_0

	b	x_0	λ	r	R^2
		$x_0 = 0$			
ZT Gao method	2.19	0.00	5.37	0.99482	0.99259
Confidence interval	(1.94, 2.22)	(0.00, 0.59)	(4.51, 5.55)	—	—
Analysis methods	3.53	−2.30	7.81	0.98528	0.99247
Gaussian distribution	—	—	—	—	0.99194
		$x_0 = 1$			
ZT GAO method	2.25	0.91	5.46	0.99489	0.99293
Confidence interval	(2.22, 2.56)	(0.00, 1.59)	(5.61, 5.55)	—	—
Analysis methods	3.53	−1.30	7.81	0.98528	0.99247
Gaussian distribution	—	—	—	—	0.99194
		$x_0 = 5$			
ZT GAO method	2.25	4.91	5.46	0.99489	0.99293
Confidence Interval	(2.22, 2.64)	(3.86, 5.59)	(5.55, 5.76)	—	—
Analysis methods	3.53	2.70	7.81	0.98528	0.99247
Gaussian Distribution	---	—	—	—	0.99194

Note: r is the correlation coefficient of fitting the sample data in logarithmic coordinates; R^2 represents the coefficient of determination for the ideal reliability fit. Subsequent tables are the same, so the same notes are not given.

In the case of $x_0 = 5$, the interesting thing is that the shape factor $b = 2.253$ and the scale factor $= 2.46$ obtained by the same method as $x_0 = 1$ is the same, and the safe life $N_0 = 4.91$ is 4 more than that of $x_0 = 1$, which is exactly equal to the change of x_0, $5 - 1 = 4$. This is exactly the case of "translation". So, it turns out that people think x_0 reflects the translation is not wrong but cannot be zeroed out.

10.2.2.2 Experiment 2: Change Only the Random Seed G

G represents the "seed" of random numbers. The seed changes, which means that the sample has changed, and other conditions are unchanged, that is, fixed at this time, $NN = 50$, $b = 2.5$, $\lambda = 5$, $x_0 = 1$. And changing the case of $G = 0, 1,$ and 5, the results are as follows:

In the case of $G = 0$, it can be seen that the shape parameter $b = 2.271$ obtained by the ZT Gao method falls within its confidence interval [2.129, 2.655], and this interval also includes the true population shape parameter $b = 2.5$. Similarly, the safe life $x_0 = 5.27$ is within its confidence interval [4.41, 5.90], which also includes the true population safe life $x_0 = 5$. However, the scale parameter does not match as closely. The parameters obtained by the analytical method are far from the corresponding population parameters. However, in the case of fatigue life, the determination coefficient for fitting the ideal reliability is larger for the analytical method compared to the ZT Gao method, although the relative difference is in the thousandths place. This confirms what was mentioned earlier, that relying solely on the determination coefficient to judge the superiority or inferiority of an algorithm is problematic.

Although the determination coefficient of the analytic method is larger than that of the ZT Gao method, we still approve the ZT Gao method. Why? Because the correlation coefficients obtained by the ZT Gao method are much larger than those obtained by the analytical method from the perspective of logarithmic fatigue life. In other words, when the determination coefficient is quite close, it depends on how much the correlation coefficient differs (Figure 10.2 and Table 10.2).

In the case of $G = 1$, both the relative coefficients and the coefficients of determination, as well as the estimation of the three parameters of the WD, are better in the ZT Gao method than in the analytical method.

In the case of $G = 5$, the correlation coefficient and determination coefficient obtained by the analytical method and the Zhentong Gao method are almost indistinguishable, making it difficult to choose between them. In fact, the parameters obtained by each method are quite similar.

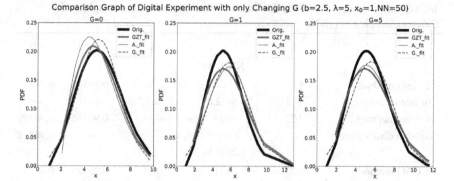

FIGURE 10.2 Comparison plot of three different fits when only changing the Random Seed G when taking random samples from the Weibull distribution.

TABLE 10.2 Comparison of Fits for Weibull Distribution Random Samples with Different Random Seeds (G)

	b	x_0	λ	r	R^2
		G = 0			
ZT GAO method	2.27	1.27	4.49	0.99288	0.98318
Confidence interval	(2.13, 2.66)	(0.41, 1.90}	(4.50, 4.73)	—	—
Analysis methods	1.86	1.98	3.65	0.96465	0.98657
Gaussian distribution	—	—	—	—	0.97720
		G = 1			
ZT GAO method	2.25	0.91	5.46	0.99489	0.99293
Confidence interval	(2.22, 2.56)	(0.00, 1.59)	(5.61, 5.55)	—	—
Analysis methods	3.53	−1.30	7.81	0.98528	0.99247
Gaussian distribution	—	—	—	—	0.99194
		G = 5			
ZT GAO method	2.18	0.90	5.33	0.99379	0.98831
Confidence interval	(1.92, 2.50)	(0.00, 1.73)	(5.34, 5.47)	—	—
Analysis methods	2.35	0.77	5.44	0.99370	0.99063
Gaussian distribution	—	—	—	—	0.99171

10.2.2.3 Experiment 3: Change Only the Sample Size NN

According to general practice, the capacity of 50 can be regarded as a "large sample". Now, fix b = 2.5, λ = 5, x_0 = 1, G = 1 and change NN = 40, 50, and 60, the result is:

From Table 10.3, it can be seen that both the correlation coefficient and the determination coefficient by the ZT Gao method are better than the analytical method at this time. As the sample size increases, the results

TABLE 10.3   Comparison of Fits for Weibull Distribution Random Samples with Different Sample Sizes (NN)

	b	x_0	λ	r	R^2
		NN = 40			
ZT GAO method	1.94	1.50	4.92	0.99365	0.99308
Confidence interval	(1.91, 2.25)	(0.64, 2.09)	(4.93, 5.27)	—	—
Analysis methods	2.77	0.04	6.49	0.98536	0.99159
Gaussian distribution	—	—	—	—	0.98832
		NN = 50			
ZT GAO method	2.25	0.91	5.46	0.99489	0.99293
Confidence interval	(2.22, 2.56)	(0.00, 1.59)	(5.61, 5.55)	—	—
Analysis methods	3.53	−1.30	7.81	0.98528	0.99247
Gaussian distribution	—	—	—	—	0.99194
		NN = 60			
ZT GAO method	2.44	0.83	5.65	0.99575	0.99604
Confidence interval	(2.37, 2.7)	(0.00, 1.50)	(5.58, 5.80)	—	—
Analysis methods	2.74	0.37	6.13	0.99478	0.99595
Gaussian distribution	—	—	—	—	0.99375

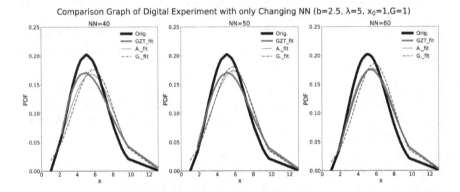

FIGURE 10.3   Comparison plot of three different fits when only changing the Sample Size NN when taking random samples from the Weibull distribution.

of the ZT Gao method are closer to the parameters of the sample, that is, the effect is better. And the estimated λ also falls within the confidence interval (Figure 10.3).

10.2.2.4 Experiment 4: Change Only the Shape Parameters b

Keeping NN = 50, λ = 5, x_0 = 1, and G = 1, we change the shape parameter b to 1.5, 2.5, 3.5, and 4.5. The results are as follows:

TABLE 10.4   Comparison of Fits for Weibull Distribution Random Samples with Different Shape Parameters (b)

	b	x_0	λ	r	R^2
		b = 1.5			
ZT GAO method	1.34	0.96	5.68	0.99495	0.99295
Confidence interval	(1.39, 1.48)	(0.61, 1.19)	(4.75, 6.85)	—	—
Analysis methods	1.87	−0.95	7.92	0.96986	0.98539
Gaussian distribution	—	—	—	—	0.97101
		b = 3.5			
ZT GAO method	3.16	0.88	5.39	0.99487	0.99288
Confidence interval	(2.79, 3.47)	(0.00, 1.99)	(4.80, 5.69)	—	—
Analysis methods	5.95	−2.52	8.85	0.98770	0.99216
Gaussian distribution	—	—	—	—	0.99190
		b = 4.5			
ZT GAO method	4.09	0.85	5.35	0.99485	0.99285
Confidence interval	(3.25, 4.40)	(0.00, 2.34)	(4.26, 5.75)	—	—
Analysis methods	9.80	−4.68	10.93	0.98829	0.99127
Gaussian distribution	—	—	—	—	0.99025

Note: The sub-table of b = 2.5 is the same as the second sub-table (x_0 = 1) of Table 1 and will not be given here.

From Table 10.4, it can be observed that both correlation coefficients and determination coefficients of the ZT Gao method are larger than those obtained by the analytical method. Therefore, we still need to take the results of the ZT Gao method. Moreover, the results of shape parameter b, safe life x0, and the confidence interval of the scale parameter λ include the population corresponding parameters. It is also seen that both is obtained by the ZT Gao method and the analytical method. The determination coefficients obtained by the Gaussian distribution are very close to each other, i.e., they fit almost "equally well", but only the results of the ZT Gao method are consistent with the population. This is also the fundamental reason why in most cases we have to use the ZT Gao method to estimate the three parameters of the Weibull distribution, because the results obtained by it are in fact better than the analytical method (Figure 10.4).

10.2.2.4 Experiment 5: Change Only the Scale Parameter λ

Keeping NN = 50, b = 2.5, x_0 = 5, and G = 1, we change the scale parameter λ to 2, 5, and 10. The results are as follows:

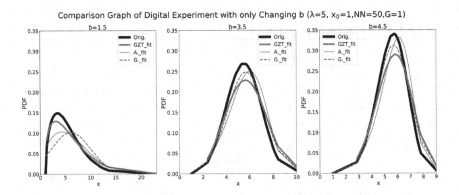

FIGURE 10.4   Comparison plot of three different fits when only changing the shape parameters b when taking random samples from the Weibull distribution.

It can be seen that when the scale parameter changes, the shape parameter, the correlation coefficient, and the determination coefficient have not changed. In this sense, the scale coefficient does only affect the proportion of the Weibull distribution. At the same time, it can be seen that the parameters obtained by the ZT Gao method are closer to the original parameters of the sample than those obtained by the analytical method, and the correlation coefficient and the coefficient of determination are larger than those obtained by the analytical method. As in the case of Experiment 1, that is, the coefficient of determination obtained by the ZT Gao method is the largest among the three methods, and the estimated values of the three parameters obtained are also the closest to the true value (Figure 10.5 and Table 10.5).

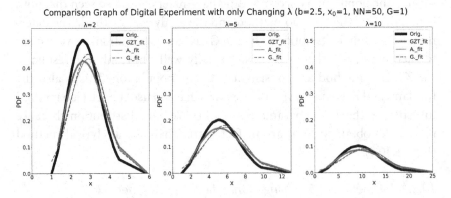

FIGURE 10.5   Comparison plot of three different fits when only changing the scale parameter λ when taking random samples from the Weibull distribution.

TABLE 10.5 Comparison of Fits for Weibull Distribution Random Samples with Different Scale Parameters (λ)

	b	x_0	λ	r	R^2
		$\lambda = 2$			
ZT GAO method	2.26	0.96	2.19	0.99489	0.99296
Confidence interval	(2.20, 2.63)	(0.55, 1.24)	(2.30, 2.22)	—	—
Analysis methods	3.53	0.08	3.12	0.98528	0.99247
Gaussian distribution	—	—	—	—	0.99194
		$\lambda = 5$			
ZT GAO method	2.25	0.91	5.46	0.99489	0.99293
Confidence interval	(2.22, 2.56)	(0.00, 1.59)	(5.61, 5.55)	—	—
Analysis methods	3.53	−1.30	7.81	0.98528	0.99247
Gaussian distribution	—	—	—	—	0.99194
		$\lambda = 10$			
ZT GAO method	2.25	0.82	10.93	0.99489	0.99293
Confidence interval	(2.22, 2.26)	(0.00, 2.18)	(10.13, 11.09)	—	—
Analysis methods	3.53	−3.61	15.61	0.98528	0.99247
Gaussian distribution	—	—	—	—	0.99194

10.3 DETERMINE THREE PARAMETERS AND CONFIDENCE INTERVAL OF WEIBULL DISTRIBUTION BY THE BOOTSTRAP

10.3.1 Digital-Experiment using the Bootstrap

The Bootstrap was researched and developed by Efron in 1997 (Efron & Hastie, 2019). The basic idea is as follows. Assuming that there are n observations to obtain the regression parameter b^, how to estimate the accuracy of this regression parameter? The Bootstrap method randomly selects n samples (called the Bootstrap samples) in these n samples with replacement. Obviously, there are repeated selections in the selected samples, and there are also unselected observations, but still regression parameter estimates can be obtained using the original procedure. Such Bootstrap calculation can be performed m times, and m the Bootstrap-estimated regression parameters b^ can be obtained. Their mean value b*^ can be used as the estimation of parameter b, and its standard deviation is recorded as s(b*^), which is called the Bootstrap standard deviation and is an estimate of the standard deviation of the sampling distribution of b^, which gives the precision of the regression parameter b. This so-called Bootstrap method is essentially a Digital-Experiment that can be extended to determine the three parameters and their intervals of

Weibull distribution, although these parameters are not really "regression parameters".

10.3.1.1 Experiment 6: The Bootstrap for Small Samples

Using the data from Example 8.2 and the Python code (refer to Code C in the book), the following results were obtained:

It can be seen from Table 10.6 that in the case of small samples, the results obtained by the non-Bootstrap method (i.e., directly using the ZT Gao method) are better than the Bootstrap method except for the scale confidence interval. The advantage of the Bootstrap method is that it simultaneously gives confidence intervals for all three parameters of the Weibull distribution, albeit quite large. It can also be seen from the histogram that it is not a strict Gaussian distribution, so the standard deviation is still relatively large. Of course, this is related to the small sample size. At the same time, it would not certainly be seen that the more m the better, the fact is that the effect of m = 1,000 is better than that of m = 1,500 (Figure 10.6).

10.3.1.2 Experiment 7: The Bootstrap for Large Samples

Using the data from Example 9.4, the following results were obtained:

From Table 10.7, it can be seen that for the so-called large sample, the Bootstrap method has little to do with the number of extractions. The values of the three parameters of the Weibull distribution and the coefficient of determination obtained by the three different times of the Bootstrap method are almost indistinguishable. In other words, 500 draws are good enough. Moreover, the estimated values of the three parameters obtained are quite close to those obtained by the ZT Gao

TABLE 10.6   Comparison of Estimated Values of WD's Three Parameters from Small Samples the Bootstrap and non-Bootstrap

	b	N_0	λ	r	R^2
ZT GAO method	2.04	277.00	320.55	0.99922	0.99823
Confidence interval	(1.99, 2.80)	(159.0, 333.0)	(320.6, 343.3)	—	—
The Bootstrap (500 times)	2.37	255.68	340.58	0.98898	0.99702
Confidence interval	(0.28, 4.47)	(79.0, 432.4)	(128.4, 552.8)	—	—
The Bootstrap (1,000 times)	2.37	254.02	342.67	0.98927	0.99728
Confidence interval	(0.30, 4.44)	(72.5, 435.6)	(124.2, 561.2)	—	—
The Bootstrap (1,500 times)	2.36	255.27	341.41	0.96666	0.99726
Confidence interval	(0.28, 4.44)	(72.7, 437.9)	(122.9, 559.9)	—	—

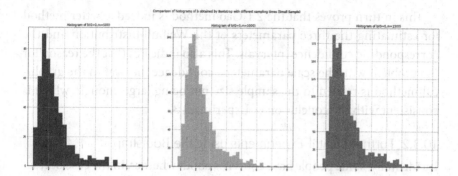

FIGURE 10.6 Comparison chart of histograms of shape parameter b obtained from Bootstrap with different sampling times for small samples.

TABLE 10.7 Comparison of Estimated Values of WD's Three Parameters from Large Samples the Bootstrap and non-Bootstrap

	b	N_0	λ	r	R^2
ZT GAO method	2.147	2.780	2.867	0.99376	0.98903
Confidence interval	(2.2, 2.27)	(2.5, 3.0)	(2.8, 2.9)	—	—
The Bootstrap (500 times)	2.160	2.801	2.844	0.97162	0.98996
Confidence interval	(1.59, 2.73)	(2.4, 3.2)	(2.4, 3.3)	—	—
The Bootstrap (1,000 times)	2.159	2.795	2.849	0.97939	0.98982
Confidence interval	(1.59, 2.73)	(2.4, 3.2)	(2.4, 3.3)	—	—
The Bootstrap (1,500 times)	2.158	2.795	2.852	0.98735	0.98970
Confidence interval	(1.60, 2.72)	(2.4, 3.2)	(2.4, 3.3)	—	—

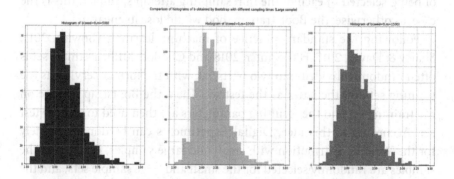

FIGURE 10.7 Comparison chart of histograms of shape parameter b obtained from Bootstrap with different sampling times for large samples.

method, and the coefficient of determination is a little better than that of the ZT Gao method, but the confidence interval obtained is a little worse than that obtained by the ZT Gao method (Figure 10.7).

This in turn proves that the ZT Gao method is indeed a better method for estimating the three parameters of the Weibull distribution and the corresponding confidence intervals. This also indicates that the reason for the problem with the scale parameter's confidence interval in the graphical method is related to the sample size not being large enough, which is consistent with the conclusion of Experiment 3.

10.3.2 Further Digital-Experiments using the Bootstrap

Incidentally, some people have concerns about whether, in the case of resampling samples with the same capacity, some samples might be repeatedly drawn while others may never be drawn. They assume that with a sample capacity of n, the probability of a sample being selected once is $1/n$, and the probability of not being selected is $1 - 1/n$. The probability of not being selected in n consecutive resampling attempts is $(1 - 1/n)^n \rightarrow e^{-1} = 0.368$ (as $n \rightarrow \infty$). This means that when using the Bootstrap method, there is a 36.8% chance that samples will never be drawn if the number of resampling attempts is sufficiently large.[1] Unfortunately, this assumption is incorrect and misleading. Since sampling is random, each sampling is a random event, and therefore, it is not possible for each sample to be unselected in every resampling attempt. The probability calculated above represents the probability that each sample is not selected in each of the m resampling attempts, rather than the probability of "never being drawn". In this sense, the Bootstrap method's sampling is fair, with a 63.2% chance of being selected in each of the m resampling attempts. In fact, this is the case, as otherwise, the Bootstrap method would lose its meaning.

Following the standard practice of machine learning (Bowles, 2017, Shalev & David, 2017, Haslwanter, 2018 and Chollet, 2018), a sample set is often randomly divided into two parts 7:3, with the former serving as the training set and the latter as the test set. Then, the Bootstrap is done on the training set and the resulting parameters are then used to test the test set. According to this idea, Digital-Experiments can be done separately for the Weibull distribution with small and large samples, i.e., 70% of the samples are randomly selected for the small and large sample sets, and the coefficients of determination of the fit of the Weibull distribution with the remaining 30% of the test set are compared without the Bootstrap and with the Bootstrap (various Digital-Experiments can be done by taking various initial variables). And this method can be used to solve exactly the problem raised in the previous section, i.e., it can be used to compare which set of parameters is more appropriate. Even if the one that fits the test set better (larger determination coefficient) should be more appropriate, it also indicates a better predictive power.

The data set can be divided into two sets of 7:3 training and test using a randomized method. Directly use the ZT Gao method and the Bootstrap to calculate the three parameters of the Weibull distribution for the training set, respectively, and then fit the test set to see which one has a larger coefficient of determination.

10.3.2.1 Experiment 8: Machine Learning with Different Methods for Small Samples

The results for small sample sets (using the data from Experiment 6) are as follows:

```
N = [350, 380, 400, 430, 450, 470, 480, 500, 520, 540, 550, 570, 600, 610, 630, 650, 670, 730,
770, 840]
Random Seed, 0, α = 0.05
Train Set is, [380, 400, 450, 470, 480, 500, 520, 540, 550, 610, 630, 730, 770, 840]
Test set is, [350, 430, 570, 600, 650, 670]
```

This is an interesting result that the three parameters of the Weibull distribution obtained by the Bootstrap from the training set appear to be closer to the results of Example 8.2, but the determination coefficients are not as large as those obtained directly from the training set by the ZT Gao method. However, the situation changes when fitting the test set, i.e., the parameters obtained by the Bootstrap are larger than the determination coefficients obtained by the ZT Gao method, so in this sense, the parameters obtained by the Bootstrap can be considered better than those obtained directly by the ZT Gao method. At this point, because the test set data is only 14, it is likely that the phenomenon of overfitting has occurred (Bowles, 2017), so that neither the direct use of the ZT Gao method nor the use of the Bootstrap gives more desirable results. In a sense, it is not so feasible to determine the three parameters and confidence intervals of the Weibull distribution using machine learning. Of course, one can continue to do Digital-Experiments on the Bootstrap, but the results are not very good and the results are not given here. It is better to look at the case of large samples (Figure 10.8 and Table 10.8).

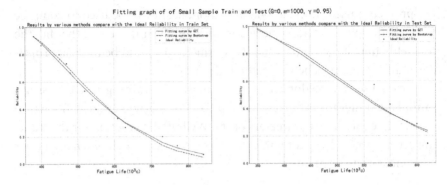

FIGURE 10.8   Fitting curve of small sample training set and test set (random seed G = 0, m = 1,000, γ = 0.95).

TABLE 10.8   Small Sample Training Set and Test Set Obtained by Different Methods

	b	x_0	λ	r	R² (Train)	R² (Test)
ZT GAO method	1.50	335	262	0.99245	0.98470	0.83839
Confidence interval	(1.81, 2.11)	(280.0, 349.0)	(230.1, 335.9)	—	—	—
The Bootstrap (m = 500)	1.79	316	283	0.99044	0.97922	0.84940
Confidence interval	(1.73, 1.86)	(310.2, 321.3)	(275.1, 291.1)	—	—	—
The Bootstrap (m = 1,000)	1.83	313	286	0.99007	0.97789	0.85031
Confidence interval	(1.78, 1.88)	(309.3, 317.5)	(280.1, 291.5)	—	—	—

10.3.2.2 Experiment 9: Machine Learning with Different Methods for Large Samples

The results for large samples (using the data from Experiment 7) are as follows:

```
Random Seed = 0, alpha = 0.05
Train Set is, [3.08, 3.26, 3.32, 3.48, 3.49, 3.56, 3.69, 3.7, 3.78, 3.8, 3.87, 4.07,
4.08, 4.1, 4.12, 4.2, 4.24, 4.31, 4.36, 4.54, 4.6, 4.62, 4.63, 4.67, 4.67, 4.72, 4.73,
4.75, 4.77, 4.84, 4.92, 4.93, 4.95, 4.96, 4.99, 5.06, 5.06, 5.1, 5.12, 5.15, 5.18, 5.2,
5.22, 5.38, 5.46, 5.47, 5.53, 5.56, 5.6, 5.64, 5.68, 5.82, 5.91, 5.94, 5.95, 5.99,
6.08, 6.13, 6.16, 6.21, 6.32, 6.33, 6.36, 6.81, 7.0, 7.35, 7.82, 7.88, 8.31, 9.87]
Test Set is, [3.79, 3.95, 4.25, 4.28, 4.31, 4.58, 4.65, 4.8, 4.82, 4.9, 4.98, 5.02, 5.03,
5.08, 5.41, 5.61, 5.63, 5.65, 5.69, 5.73, 5.86, 6.04, 6.19, 6.26, 6.41, 6.46, 7.96, 8.45,
8.47, 8.79]
```

The results still appear to be somewhat different from the small sample case, i.e., the Bootstrap obtained in the training set for the three parameters of the Weibull distribution than directly with the ZT Gao method looks better, and the coefficient of determination is also better than the ZT Gao method. The results for the coefficients of determination in the test set are still the same, so in this sense, it can still be said that the Bootstrap is better than the direct ZT Gao method. In fact, it is understandable that the Bootstrap is better than the direct use of the ZT Gao method. This is because at this time, the training set used for both the direct and the Bootstrap is 70% of the original data set, but the direct method can only be used once, while the Bootstrap can be repeated 1,000 times. So on average, the Bootstrap should have better results, although the fitted graphs are almost indistinguishable, but the fitted determination coefficients are still distinguishable from the size, although the difference is only to the third decimal place (Figure 10.9 and Table 10.9).

10.3.2.3 Experiment 9: Machine Learning with Different Methods for Large Samples (Changed Random Seed)

We can still conduct Digital-Experiments similar to Section 10.2, such as changing the random seed from 0 to 300, and the running results are as follows:

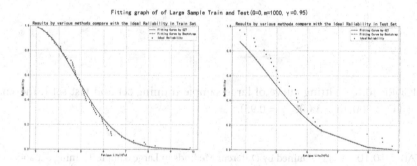

FIGURE 10.9　Fitting curve of large sample training set and test set (random seed G = 0, m = 1,000, γ = 0.95).

TABLE 10.9　Large Sample Training Set and Test Set Obtained by Different Methods.

	b	x_0	λ	r	R^2 (Tran Set)	R^2 (Test Set)
ZT GAO method	2.048	2.75	2.75	0.99485	0.99072	0.97506
Confidence interval	(2.27, 2.05)	(2.40, 3.00)	(2.75, 2.87)	—	—	—
The Bootstrap (m = 500)	2.130	2.73	2.78	0.99636	0.99210	0.97876
Confidence interval	(2.09, 2.17)	(2.71, 2.75)	(2.75, 2.81)	—	—	—
The Bootstrap (m = 1000)	2.132	2.72	2.78	0.99641	0.99219	0.97850
Confidence interval	(2.11, 2.16)	(2.71, 2.74)	(2.76, 2.80)	—	—	—

```
Random Seed = 300, alpha = 0.05
Train Set is, [3.08, 3.26, 3.32, 3.49, 3.56, 3.69, 3.7, 3.78, 3.8, 3.87, 3.95, 4.07, 4.08,
4.1, 4.2, 4.24, 4.28, 4.36, 4.58, 4.6, 4.62, 4.65, 4.67, 4.67, 4.72, 4.73, 4.75, 4.77,
4.82, 4.84, 4.9, 4.92, 4.95, 4.96, 4.99, 5.02, 5.06, 5.06, 5.12, 5.15, 5.18, 5.2, 5.22,
5.41, 5.53, 5.56, 5.6, 5.63, 5.64, 5.65, 5.68, 5.69, 5.73, 5.86, 5.95, 6.04, 6.08, 6.13,
6.16, 6.21, 6.26, 6.36, 6.41, 7.0, 7.82, 7.88, 7.96, 8.31, 8.79, 9.87]
Test Set is, [3.48, 3.79, 4.12, 4.25, 4.31, 4.31, 4.54, 4.63, 4.8, 4.93, 4.98, 5.03, 5.08,
5.1, 5.38, 5.46, 5.47, 5.61, 5.82, 5.91, 5.94, 5.99, 6.19, 6.32, 6.33, 6.46, 6.81, 7.35,
8.45, 8.47]
```

This result indicates that the random seeds have been changed, i.e., the training set and the test set have been changed, but the conclusion remains the same. The determination coefficients obtained by the Bootstrap "beat" the direct use of the ZT Gao method both in the training and test sets. Moreover, the determination coefficient obtained in

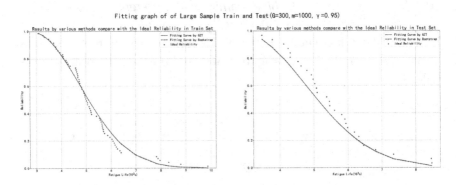

FIGURE 10.10   Fitting curve of large sample training set and test set (random seed G = 300, m = 1,000, γ = 0.95).

TABLE 10.10   Data Obtained by Different Methods in Large Sample Training Set and Test Set (Changed Random Seed)

	b	x_0	λ	r	R^2 (Train Set)	R^2 (Test Set)
ZT GAO method	2.048	2.75	2.81	0.99182	0.98392	0.98267
Confidence interval	(2.14, 2.06)	(2.50, 3.00)	(2.68, 2.93)	—	—	—
The Bootstrap (m = 500)	2.077	2.75	2.80	0.99317	0.98510	0.98419
Confidence interval	(2.05, 2.11)	(2.73, 2.77)	(2.78, 2.83)	—	—	—
The Bootstrap (m = 1,000)	2.073	2.76	2.80	0.99317	0.98510	0.98412
Confidence interval	(2.05, 2.09)	(2.74, 2.77)	(2.78, 2.82)	—	—	—

the test set is much larger than that obtained when the random seed is 0 (Figure 10.10 and Table 10.10).

Such Digital-Experiments can be continued, such as taking other experimental data, etc. However, these can be used as exercises for readers to get a better grasp of the Bootstrap as well as the Digital-Experimental method.

NOTE

1 Refer to https://zh.wikipedia.org/wiki/%E8%87%AA%E5%8A%A9%E6% B3%95

REFERENCES

Bowles M (2017), *Machine Learning in Python: Essential Techniques for Predictive Analysis*, Posts & Telecom Press, Beijing.

Chollet F (2018), *Deep Learning with Python*, Posts & Telecom Press, Beijing.

Efron B & Hastie T (2019), *Computer Age Statistical Inference*, China Machine Press, Beijing, pp. 117–119.

Haslwanter T (2018), *An Application to Statistics with Python*, Posts & Telecom Press.

Li SZ (2004), *Mathematical Experiment*, 2nd, Higher Education Press, Beijing.

Montgomery DC, Peck EA, & Geoffrey Vining G (2019), *Introduction to Linear Regression Analysis*, 4th, China Machine Press, Beijing, pp. 377–380.

Shalev SS & David SB (2017), *Understanding Machine Learning*, China Machine Press, Beijing.

Wang YT (1988), Digital Wind Tunnel, *Aerodynamic Missile Journal*, 12: 56–60.

Xu JJ (2022), Digital Experiment for Estimating Three Parameters and Their Confidence Intervals of Weibull Distribution, *International Journal of Science, Technology and Society*, 10(2): 72–81. doi: 10.11648/j.ijsts.20221002.16

Epilogue

Perspectives on Intelligent Fatigue Statistics

WHAT IS THE DIFFERENCE between AI (artificial intelligence) and intelligence? Generally speaking, "intelligence" is a characteristic of human beings, so whether it is intelligence or artificial intelligence, it means that people make something "intelligent". For example, the book emphasizes that the "intelligence" in "Intelligent Fatigue Statistics" is "embodied in the 'no-table' (no more numerical tables of statistical functions will be attached after the book), and except for the sample data that requires manual input, all the results of data processing, including graphing, are done automatically by the computer".

Further examples can be used to illustrate this. A car navigation system, for example, can give one a route from A to B, but it does not guarantee that it is an optimal route, nor does it tell you whether you will encounter traffic jams on the road. At this time, it can only be said to be in a primary stage of intelligent navigation. Wait until it can tell you that the route it chooses is "optimal", or at least the least congested, and can adjust in real time. Then, it can be said that navigation AI is better than humans themselves, and an advanced state of AI has been reached. There is no need to make a fuss about it, in the same way people can't catch up with cars no matter how fast they run. Similarly, in the third part of this book, the computer replaces almost all the manual calculations and replaces the human working diagram, especially with the ZT Gao method, so that people can easily get the estimated value of the three parameters of Weibull distribution, the embodiment of the intelligence of "fatigue statistics", but it is far from the automatic recording of various fatigue-related real-time data to give the component life estimate and remind people in real time of the degree of risk that may occur and suggest measures to take. In other words, it has reached the level of "intelligent structure" advocated by Mr. Gao, which is probably the advanced stage of "Intelligent Fatigue Statistics" or the advanced stage of AI application in this area.

The mathematical basis of intelligence or AI is certainly not only mathematical inference, but mathematical inference is undoubtedly a very important part of it. How important is it? Without new algorithms based on mathematical inference, such as deep learning based on neural networks, modern AI would not be possible. These new algorithms are not used in "Intelligent Fatigue Statistics", except for Bootstrap. However, it is not difficult to find that the parameter estimation of statistical distributions is essentially a problem of finding probability extremum. Therefore, any optimization algorithm in AI can be used, such as the nonlinear programming method (Avriel, 1976), Boltzmann annealing method (Zheng, 2015), etc. Recently, the second author of this book has also made beneficial attempts in this area and achieved preliminary results (Xu, 2023A).

Now, we would like to present three possible scenarios for the further application of AI in "Intelligent Fatigue Statistics".

1. The most interesting scenario for the authors is the "digital fatigue machine" proposed in Chapter 10, which is very difficult to produce, but if it is manufactured, its benefits will not be lower than those of the "digital wind tunnel".

2. Another scenario calls for directly obtaining the fatigue load spectrum. As pointed out in Gao & Xiong (2000), "to ensure structural safety and reliability, the first problem is to accurately predict the service life of the structure. And the prerequisite for solving the fixed life is to prepare fatigue load and its environmental spectrum that simulates the actual working condition". So how can these spectra be prepared? It mainly relies on the accumulation of relevant data, which is then classified and simplified according to different criteria. There is also the problem of extrapolating the population distribution from the sample. Of course, the data at this time is quite "much" ("big data") and complex (the geometric structure itself is complex,

and random factors are also considerable), so to find a more realistic load spectrum is definitely not an easy task. This time, the intervention of AI is a very important part of the process and will make this difficult task easy to handle. One only needs to input the initial conditions such as the geometry of the structure and the working environment to get a load spectrum that is more in line with the actual situation.

3. Finally, the use of AI to directly derive the schedule of major and minor repairs and obsolescence would be extremely beneficial. The most important indicator of the life of an aircraft is "flight hours", and every aircraft has a schedule of major and minor repairs that is based on the theory of "fatigue reliability", but the process of getting this schedule is very complicated. We hope that in the future, we can get such a schedule automatically by using AI technology as long as the original data has been put in.

Noted here are just three possible uses of AI in "fatigue statistics", and we believe there are more application scenarios. Thus, the authors hope that readers can creatively solve various problems in their work based on the theories and methods learned in this book, especially in regard to "Intelligent Fatigue Statistics" and further contribute to the application of more advanced AI. We believe that there is indeed great potential in this field.

REFERENCES

Avriel J (1976), *Nonlinear Programming Analysis and Methods,* Shanghai Science and Technology Press, Shanghai.

Gao ZT & Xiong JJ (2000), *Fatigue Reliability*, Beihang University Press, Beijing, p. 331.

Xu JJ (2023A), From Zhentong Gao Method to Generalized Zhentong Gao Method, *Journal of Physics: Conference Series(Vol. 2650)*, doi:10.1088/1742-6596/2650/1/012013

Zheng J (2015), *Machine Learning*, Electronic Industry Press, Beijing, pp. 216–219.

Postscript

THE CHINESE EDITION OF *Intelligent Fatigue Statistics (IFS)*, co-authored by Mr. Gao Zhentong, an academician of the Chinese Academy of Sciences, and myself, was published in October 2022 in Beijing. This book is, in fact, a crystallization of a "60-year teacher-student relationship" between Mr. Gao and myself. Many stories can be found in it, but due to space limitations here, I can only choose one or two to share with readers.

Sixty years ago, in 1963, when I was 17 years old, from China southern city Guangzhou, I entered the Department of Aeronautical Mathematics and Mechanics of the Beijing Aviation Institute in Beijing. At that time, Mr. Gao was the part-time deputy director of the department and also the main instructor of the Material Mechanics.

At the beginning of 1966, at the end of the school's unified examination of Material Mechanics, Mr. Gao gave me the only perfect grade of 100, which led to some people suspecting that there was a personal relationship between us, and this matter created a lot of noise, and consequently, during the "Cultural Revolution", Mr. Gao had to withstand a deal of pressure for it. It became our "negative assets". After the so-called "graduating" in 1968, I was assigned to the Heilongjiang Military Farm.

In 1978, after the end of the Cultural Revolution, I said goodbye to Heilongjiang and returned to Guangzhou to go to graduate school. Not wanting to go back to the north, I went to Sun Yat-sen University to study in the Department of Physics.

In October 2012, for Beihang University's 60th anniversary, Mr. Gao published an article, "Forty Years of Teacher-Student Love", in *Beihang University* News. He wrote that the things that had existed in the family for decades "have been lost for various reasons, but I still have the letters from students. One of my most valuable treasures is a 46-year-old test paper of Xu Jiajin's 'Summary of the Energy Law' in 1966. Why do I keep this paper? Because he answered this exam paper very well and had unique insights ...... so I gave him 100 points ...... I don't have any other collectibles; this is my best collection".

When I saw the text, I felt very shocked and ashamed for I really had not appreciated Mr. Gao's good intentions and glorification. From then on, I secretly resolved to try to repay Mr. Gao's kindness.

I have always been interested in mathematics, mechanics, and physics, I am especially fascinated by computers, and almost all of my papers published in domestic and international magazines are related to scientific computing. In recent years, with artificial intelligence (AI) developing rapidly, I began to think that AI should have a role to play in "fatigue statistics". Therefore, I studied Mr. Gao's works in this area, carefully read his nearly 400-page monograph, and making more than 40,000 words of reading notes, re-calculated all the example problems using Excel and Python. I felt it was time to pay back Mr. Gao.

In November 2019, I returned to my alma mater from Guangzhou to "worship the master for the second time" and became his "closed disciple of the fatigue portal". Mr. Gao instructed me to write *Intelligent Fatigue Statistics*. Taking nearly 3 years, it was revised more than ten times before being finalized by Mr. Gao. Professor Bao Rui, Mr. Gao's "little disciple", wrote the Foreword for this book and made irreplaceable contributions to the publication of this book in its Chinese version.

At the age of 94, Mr. Gao guided his 76-year-old student to create a new field of Intelligent Fatigue Statistics. As Mr. Gao pointed out, "The two of us are more than 2000 km apart and do not work at school, but the co-authored book of more than 270,000 words, *Intelligent Fatigue Statistics*, was actually published, and there seems to be no such precedent in ancient and modern times, which cannot but be said to be a miracle. In fact, it is not strange to say strange, blessed to the heart, so that buried for more than 40 years for the Weibull distribution of unique insights to see the light of day".

Mr. Gao suggested that I publish two or three academic papers in domestic and international journals before formally completing the book manuscript, so as to make his innovative ideas known to the public and in order to listen to the opinions of my peers. Despite many twists and turns, the three papers were published in 2021 and 2022 as scheduled. In any case, this is something that the over 90-year-old Mr. Gao felt particularly gratified by because he has always believed that the Weibull distribution is a more general full-state distribution than the Gaussian distribution, especially in the statistical extrapolation of the fatigue life obtained at a 100% safe life. This theory of structural reliability is particularly important, but due to the complexity of Weibull distribution, the estimation of its three parameters is made even more difficult. This is particularly important for structural reliability theory, and the complexity of Weibull distribution and the difficulty in estimating its three parameters have affected its wide application.

I put forward the "Zhentong Gao method" using computer intelligence to solve the problem of estimating three parameters of Weibull, giving 100% reliability of safe life, and greatly promoting a wider range of applications for fatigue statistics intelligence.

In retrospect, while this achievement might seem to have sprouted from that score of 100 some 60 years ago, it is actually one of the fruits of the appreciation that the teacher has shown for his students.

A few months ago, I visited Mr. Gao in Beijing, and he instructed me to translate this book into English so that English readers can share the results of the book. According to his request, I have invited Dr. Yang Mingqing, an alumnus of Sun Yat Sen University, as the proofreader for the English version, as well as Mr. Ni Wanmei, an alumnus of the University of Arizona in the United States, to do the polishing work for the English manuscript. Without their participation, the English version of this book would not have been successfully published.

Jiajin Xu
Guangzhou, China

Appendix A

Python Code for Calculating Confidence Interval Curve of Weibull Distribution

```python
import numpy as np
from scipy.optimize import leastsq
import scipy.stats as stats
import matplotlib.pyplot as plt
N=[350,380,400,430,450,470,480,500, 520,540,550,570,600,610,630,650,670,730,770,840]
LS=len(N);E=1e-2;I=int(N[0]/E)
N00=[i*E for i in range(I)]
alp=0.1;gamm=1-alp;R=0.999
print('N=',N)

def fX(N0):
    x=[np.log(N[i]-N0) for i in range(LS)]
    return np.array(x)

def P_lim(n,alp):
    Pu=[]
    for i in range(1,n+1):
        Pu.append(stats.beta.isf(alp,i,n-i+1))
    PL=[1-Pu[n-i] for i in range(1,n+1)]
    return Pu,PL

def linearF(p,X,Y):
    b,d=p
    return Y-b*X-d

def maxF(N00,Y):
    r=[];b=[];lamda=[]
    for i in N00:
        X=fX(i)
        result=leastsq(linearF,[1,1],args=(X,Y))
        b1,d1=result[0]
        r1,f=stats.pearsonr(X,Y)
        r.append(r1);b.append(b1);lamda.append(np.exp(-d1/b1))
    km=r.index(max(r));bm=b[km];Lm=lamda[km];N0=N00[km];rm=r[km]
    return N0,rm,bm,Lm
```

```
def Weib(R,b,N0,Lmd):
    return Lmd*np.exp(np.log(-np.log(R))/b)+N0

Y=[1-i/(LS+1) for i in range(1,LS+1)]
y=[np.log(np.log(1/(1-i/(LS+1)))) for i in range(1,LS+1)]
Yv=np.array(y)
Yu,YL=P_lim(LS,alp)
print('In Confidence=',gamm,'and Size=',LS,': super limit of Confidence:')
for i in range(LS):print('%.2f'%(100*Yu[i]),',',end='')
print()
print('In Confidence=',gamm,',and Size=',LS,': lower limit of Confidence:')
for i in range(LS):print('%.2f'%(100*YL[i]),',',end='')
print()
Yup=[np.log(np.log(1/(1-Yu[i]))) for i in range(LS)]
YLw=[np.log(np.log(1/(1-YL[i]))) for i in range(LS)]
Yu=np.array(Yup);YL=np.array(YLw)


N0_v,r_v,b_v,L_v=maxF(N00,Yv)
print('Parameters by average rank: b=','%.3f'%b_v,',λ=', '%.2f'%L_v, ',N0=', '%.2f'%N0_v,
',r=', '%.5f'%r_v)
N0_u,r_u,b_u,L_u=maxF(N00,Yu)
print('Parameters by super rank:b=','%.3f'%b_u,',λ=','%.2f'%L_u,',N0=','%.2f'%N0_u,',r=','%.5f'%r_u)
N0_L,r_L,b_L,L_L=maxF(N00,YL)
print('Parameters by lower rank:b=','%.3f'%b_L,',λ=','%.2f'%L_L,',N0=','%.2f'%N0_L,',r=','%.5f'%r_L)
Nv=Weib(R,b_v,N0_v,L_v);Nu=Weib(R,b_u,N0_u,L_u);NL=Weib(R,b_L,N0_L,L_L);
print('Confidence Interval of b=','%.3f'%b_v,'is:[','%.3f'%b_L,',','%.3f'%b_u,']')
print('Confidence Interval of N0=','%.3f'%N0_v,'is:[','%.3f'%N0_u,',','%.3f'%N0_L,']')
print('In Reliability=',R,':Confidence Interval of fatigue Life(%.3f'%Nv,')is[%.3f'%Nu,',%.3f'%NL,']')
Yvw=[np.exp(-np.power((N[i]-N0_v)/L_v,b_v)) for i in range(LS)]
Yuw=[np.exp(-np.power((N[i]-N0_u)/L_u,b_u)) for i in range(LS)]
YLw=[np.exp(-np.power((N[i]-N0_L)/L_L,b_L)) for i in range(LS)]
b1=str('%.2f'%b_v);r1=str('%.5f'%r_v);gm=str('%.2f'%gamm)
plt.title('Figure of fatigue reliability confidence limit(γ='+gm+')')
plt.xlabel('Fatigue Life N')
plt.ylabel('Reliability')
plt.plot(N,Yvw,linewidth=3,c='r',label='Reliability by average rank')
plt.plot(N,Yuw,linewidth=2,c='b',label='Reliability by upper rank',linestyle="-.")
plt.plot(N,YLw,linewidth=2,c='g',label='Reliability by lower rank',linestyle="--")
plt.scatter(N,Y,c='k',marker='o',label='Idear Reliability')
plt.grid()
plt.legend()
plt.show()
```

Appendix B

Python Code of Digital Experiment of Weibull Distribution Parameters

```python
import numpy as np
import scipy.stats as stats
from scipy.optimize import leastsq
import math
import matplotlib.pyplot as plt
import pandas as pd
NN=50;b0=2.5;λ0=5;N0=1;G0=1
α=0.05;gamm=1-α

def main(NN, b0, λ0, N0, G0, K, wt, Str):

    def randomWN(NN, b0, λ0, N0, G0):
        np.random.seed(G0)
        NR=np.random.rand(NN)
        N=[N0+λ0*np.exp(np.log(np.log(1/i))/b0) for i in NR]
        N.sort()
        return N
    N=randomWN(NN, b0, λ0, N0, G0);
    N1=np.copy(N);N1=list(N1);N1.insert(0,N0)
    LS = len(N);E = 1e-2;I = int(N[0]/E)
    N00=[i*E for i in range(I)]
    print('Sample:b=',b0,',λ=',λ0,',  x0=',N0,',Seed=',G0,',Size=',LS,',α,=',α)
    N1=np.array(N1);s=np.std(N,ddof=1);Nav=np.average(N);Nm=np.median(N)
    YG=stats.norm.pdf(N1,Nav,s)

    def fX(N0):
        x=[np.log(N[i]-N0) for i in range(LS)]
        return np.array(x)

    def P_lim(n,alp):
        Pu=[]
        for i in range(1,n+1):
            Pu.append(stats.beta.isf(alp,i,n-i+1))
        PL=[1-Pu[n-i] for i in range(1,n+1)]
        return Pu,PL

    def linearF(p,X,Y):
        b,d=p
        return Y-b*X-d
```

```python
def maxF(N00,Y):
    r=[];b=[];λ=[]
    for i in N00:
        X=fX(i)
        result=leastsq(linearF,[1,1],args=(X,Y))
        b1,d1=result[0]
        r1,p=stats.pearsonr(X,Y)
        r.append(r1);b.append(b1);λ.append(np.exp(-d1/b1))
        km=r.index(max(r));bm=b[km];λm=λ[km];N0=N00[km];rm=r[km]
    return N0,rm,bm,λm

Y=[1-i/(LS+1) for i in range(1,LS+1)]
y=[np.log(np.log(1/(1-i/(LS+1)))) for i in range(1,LS+1)]
Yv=np.array(y)
Yu,YL=P_lim(LS,α)
Yup=[np.log(np.log(1/(1-Yu[i]))) for i in range(LS)]
YLw=[np.log(np.log(1/(1-YL[i]))) for i in range(LS)]
Yu=np.array(Yup);YL=np.array(YLw)

N0_v,r_v,b_v,λ_v=maxF(N00,Yv)
N0_u,r_u,b_u,λ_u=maxF(N00,Yu)
N0_L,r_L,b_L,λ_L=maxF(N00,YL)
PG=1-stats.norm.cdf((N-Nav)/s)
rG,p=stats.pearsonr(PG,Y)

def GZTo(N):
    D=np.log(2);e=1.0e-7;b1=0.5;b2=10
    def F(b):
        E=math.gamma(1+2/b)-np.power(math.gamma(1+1/b),2)
        F=s*np.power(E,-0.5)
        return Nav-Nm+F*(np.power(D,1/b)- math.gamma(1+1/b))
    if F(b1)*F(b2)>0: print('this solve is wrong,please try choice b1 or b2')
    def Solve(B1,B2):
        BI=(B2+B1)/2
        k=0
        while abs(F(BI))>e:
            BI=(B2+B1)/2
            if F(B1)*F(BI)>0:
                B1=BI
            else: B2=BI
            k+=1
        return BI,k
    bo,k=Solve(b1,b2)
    return bo,k

def RR(X,Y):
    X=np.array(X);Y=np.array(Y)
    Xv=np.var(X);XYv=np.var((X-Y))
    return 1-XYv/Xv

bo,k=GZTo(N)
E=math.gamma(1+2/bo)-np.power(math.gamma(1+1/bo),2)
λ=s*np.power(E,-0.5)
No0=Nav-λ*math.gamma(1+1/bo)
if No0<N[0]:
    X=fX(No0);r,p=stats.pearsonr(X,Yv)
else:print('by Analytical method connot get consistence solution as No0>N[0]')

def W(N0,b,λ,N):
    Y=b/λ*np.power((N-N0)/λ,b-1)*np.exp(-np.power((N-N0)/λ,b))
    return Y

N=np.array(N)
Y0=W(N0,b0,λ0,N);Y01=W(N0,b0,λ0,N1)
Y1=W(N0_v,b_v,λ_v,N)
Y2=W(No0,bo,λ,N)
Y1R=RR(Y1,Y0);Y2R=RR(Y2,Y0);YGR=RR(YG,Y01);

bv='%.2f'%b_v;N0v='%.2f'%N0_v;λv='%.2f'%λ_v;rv='%.5f'%r_v;Rv='%.5f'%Y1R;
bold='%.2f'%bo;N0o='%.2f'%No0;λo='%.2f'%λ;r='%.5f'%r;Ro='%.5f'%Y2R
```

```
bu='%.2f'%b_u;N0u='%.2f'%N0_u;λu='%.2f'%λ_u;bL='%.2f'%b_L
N0L='%.2f'%N0_L;λL='%.2f'%λ_L;
bs='('+bL+' , '+bu+')';N0s='('+N0u+' , '+N0L+')'; λs='('+λu+' , '+λL+')'

table=[['     ',' b ',' x0 ',' λ ',' r ',' R2 '], [' GZT Method',bv,N0v,λv,rv,Rv],
        ['Cofidence Interval',bs,N0s,λs,'---','---'], ['Aanlytical method',bold,N0o,λo,r,Ro],
        ['Gaussian Distribution','---','---','--- ','%.5f'%rG,'%.5f'%YGR]]
table=pd.DataFrame(table)
Ks=Str+str(i)
table.to_excel(wt, sheet_name=Ks,index=False)
wt.save()

KK=130+K
plt.subplot(KK)
plt.title(Ks,fontsize=30)
plt.plot(N1,Y01,linewidth=12,label='Orig.',c='k')
plt.xlabel('x',fontsize=30);plt.xlim(0,int(max(N)+0.5))
plt.plot(N,Y1,label='GZT_fit',c='r',linewidth=8)
plt.plot(N,Y2,linewidth=2,label='A._fit',c='b')
plt.plot(N1,YG,linewidth=4,label='G._fit',c='g')
plt.ylim(0,0.35);plt.ylabel('PDF')
plt.xticks(fontsize=25);plt.yticks(fontsize=25)
plt.legend(fontsize=25)

plt.figure(figsize==(36, 13),dpi=500)
fig.suptitle('Comparison Graph of Digital Experiment with only Changing '+'x$_0$'+'
(b=2.5, λ=5,NN=50,G=1)',fontsize=40)
wt = pd.ExcelWriter('x0.xlsx')
x0=[0,1,5];K=1
for i in x0:
    main(NN, b0, λ0, i, G0,K,wt,'x0=')
K+=1
```

Appendix C

Python Code for Evaluating Weibull Distribution Parameters and Confidence Intervals by the Bootstrap

```python
import numpy as np
import scipy.stats as stats
from scipy.optimize import leastsq
import matplotlib.pyplot as plt
import pandas as pd

α=0.05;gamm=1-α
N=[350,380,400,430,450,470,480,500,520,540,550,
    570,600,610,630,650,670,730,770,840]
LS=len(N);E=1;I=int(N[0]/E)
N00=[i*E for i in range(I)]

def fX(N0,N):
    x=[np.log(N[i]-N0) for i in range(LS)]
    return np.array(x)

def P_lim(n,alp):
    Pu=[]
    for i in range(1,n+1):
        Pu.append(stats.beta.isf(alp,i,n-i+1))
    PL=[1-Pu[n-i] for i in range(1,n+1)]
    return Pu,PL

def linearF(p,X,Y):
    b,d=p
    return Y-b*X-d

def maxF(N00,Y,N):
    r=[];b=[];λ=[]
    for i in N00:
        X=fX(i,N)
        result=leastsq(linearF,[1,1],args=(X,Y))
        b1,d1=result[0]
        r1,p=stats.pearsonr(X,Y)
        r.append(r1);b.append(b1);λ.append (np.exp(-d1/b1))
```

```
    km=r.index(max(r));bm=b[km];λm=λ[km];N0=N00[km]; rm=r[km]
    return N0,rm,bm,λm


Y=[1-i/(LS+1) for i in range(1,LS+1)]
y=[np.log(np.log(1/(1-i/(LS+1)))) for i in range(1,LS+1)]
Yv=np.array(y)
Yu,YL=P_lim(LS,α)
Yup=[np.log(np.log(1/(1-Yu[i]))) for i in range(LS)]
YLw=[np.log(np.log(1/(1-YL[i]))) for i in range(LS)]
Yu=np.array(Yup);YL=np.array(YLw)


N0_v,r_v,b_v,λ_v=maxF(N00,Yv,N)
N0_u,r_u,b_u,λ_u=maxF(N00,Yu,N)
N0_L,r_L,b_L,λ_L=maxF(N00,YL,N)


def RR(X,Y):
    X=np.array(X);Y=np.array(Y)
    Xv=np.var(X);XYv=np.var((X-Y))
    return 1-XYv/Xv


P=[1-i/(LS+1) for i in range(1,LS+1)]
Pw=[np.exp(-np.power((N[i]-N0_v)/λ_v,b_v)) for i in range(LS)]

bv='%.2f'%b_v;N0v='%.2f'%N0_v;λv='%.2f'%λ_v;rv='%.5f'%r_v;Rv='%.5f'%RR(P,Pw);
bu='%.2f'%b_u;N0u='%.1f'%N0_u;λu='%.1f'%λ_L;ru='%.5f'%r_u;
bL='%.2f'%b_L;N0L='%.1f'%N0_L;λL='%.1f'%λ_u;rL='%.5f'%r_L;
bs='('+bL+' ,'+bu+')';N0s='('+N0u+' ,'+N0L+')';λs='('+λu+' ,'+λL+')'


table=[['      ',' b ',' N0 ',' λ ',' r ',' R2 '],['GZT method',bv,N0v,λv,rv,Rv],
        ['Confidence Iterval', bs, N0s, λs, '---','---']]

m=500;LSS=LS;
b0=np.array([b_v,N0_v,λ_v])
print('b=%.2f'%b0[0],',N0=%.2f'%b0[1],',λ=%.2f'%b0[2])


def Bootstrap(m,table,K,G0):
    np.random.seed(G0);k=0;B=np.copy(b0)
    while k<m:
        I = np.random.choice(range(LS), size=LSS)
        N1=[N[i] for i in I]
        N1.sort()
        N01,r1,bm1,λ1,=maxF(N00,Yv,N1)
        b1=np.array([bm1,N01,λ1])
        B=np.vstack((B,b1))
        k+=1
    print('The regression parameters by Bootstrap(m=',k,',seed=',G0,'):')
    b=np.mean(B,axis=0)
    print('b=%.2f'%b[0],',N0=%.2f'%b[1],',Lamda=%.2f'%b[2])
    SEb=np.std(B,axis=0)
    t_n=stats.norm.ppf(1-α/2)
    dta=t_n*SEb
    X=fX(b[1],N1);r1,p=stats.pearsonr(X,Yv)
    Ps=[np.exp(-np.power((N[i]-b[1])/b[2],b[0])) for i in range(LS)]

    ms=str(m);Gs=str(G0)
    mt='Bootstrap('+ms+'c)';b_m='%.2f'%b[0];N0_m='%.2f'%b[1];
    λ_m='%.2f'%b[2];R2_m='%.5f'%RR(P,Ps);r_m='%.5f'%r1
    bI='('+'%.2f'%(b[0]-dta[0])+' ,  '+'%.2f'%(b[0]+dta[0])+')'
    N0I='('+'%.1f'%(b[1]-dta[1])+' ,  '+'%.1f'%(b[1]+dta[1])+')'
    λI='('+'%.1f'%(b[2]-dta[2])+' ,  '+'%.1f'%(b[2]+dta[2])+')'
    table.append([mt,b_m,N0_m,λ_m,r_m,R2_m])
    table.append(['Confidence Iterval', bI, N0I, λI, '---','---'])

    BT=B.T
    KK=130+K
    if K==1:C='b'
    elif K==2:C='r'
    else:C='g'
    plt.subplot(KK)
    n, bins, patches = plt.hist(BT[0], bins='auto', facecolor=C, alpha=0.5)
    plt.title('Histogram of b(seed='+Gs+',m='+ms+')' ,fontsize=30)
```

```
        plt.grid()
        return table

plt.figure(figsize==(36,12),dpi=300)
fig.suptitle('Comparison of Histrograms of b obtained by bootsrapt with'+'diffrent
sampleing (smallsample}',
fontsize=40)
G0=0;Gs=str(G0)
name='small('+Gs+').xlsx'
wt = pd.ExcelWriter(name)
mm=[500,1000,1500];K=1
for i in mm:
        table=Bootstrap(i, table,K,G0)
        K+=1
T=pd.DataFrame(table)
T.to_excel(wt,'small')
wt.save()
```

Index

Printed in the United States
by Baker & Taylor Publisher Services

Printed in the United States
by Baker & Taylor Publisher Services